Unnatural Acts

Critical Thinking, Skepticism, and Science Exposed!

Robert Todd Carroll

ISBN 978-1-105-90219-2

This book is dedicated to John Renish, editor of The Skeptic's Dictionary website, whose encouragement and sage advice have been very welcome this past decade.

About This Book

I wrote *Unnatural Acts* for people who want to improve their thinking, become more accurate in their beliefs and more reasonable in their actions, and who are tired of being fooled by others.

This book is about natural and unnatural thinking, and how the way we think affects everything we do. Natural thinking is instinctive, intuitive, quick and dirty. It works pretty well most of the time, but it can get us into trouble. We can deceive ourselves into believing what's not true or even what goes against our own self-interest, if we're not careful. And manipulators who understand natural thinking can use that understanding to hoodwink us into believing what isn't true or doing what they want us to do. You can reduce the chances of being duped by learning how to think in unnatural ways. I hope this book helps you do that.

On the other hand, if you think you already know everything you need to know and are absolutely certain that what you know is true, you may still find this book helpful, especially if you make your living tricking others into believing what you want them to believe.

Table of Contents

I
Believing in the Palpably Not True

"I know what I believe. I will continue to articulate what I believe and what I believe—I believe what I believe is right."— George W. Bush

"The most common of all follies is to believe passionately in the palpably not true. It is the chief occupation of mankind." —H. L. Mencken

---

Natural Illusions and the Brain at Work

H. L. Mencken did not have a very high opinion of the human species. To write, as Mencken did, that our *chief* occupation is to believe in the palpably untrue may seem to be an obvious exaggeration. We might even say that his pessimistic claim is untrue, but that he passionately believed it anyway. He painted all mankind into the same corner, there to sit with our dunce hats on, raging against the truth. It is no exaggeration, however, to describe *half* of all adults in America as passionately defending the palpably not true. I am referring to the fact that poll after poll finds that about 50% of our adult population does not accept what just about every scientist on the planet accepts: the fact that humans and all other animals have evolved by processes such as natural selection over hundreds of millions of years on a planet that revolves around a star that dwells in a remote corner of one of billions of galaxies. I know that many find this a bleak picture of the universe and blame scientists for taking away the magic of a universe created just for us by an Almighty God. While I consider the story of creation unworthy of our age, I do not consider the people who hold such a view to be any less intelligent or knowledgeable than I am. Throughout this book I will use harsh language to characterize various ideas, but my acerbic tongue is aimed at ideas and not at the people who adhere to those ideas. Many of the ideas I now consider unworthy of belief are ideas that I once accepted as truths, including the idea that the universe was designed and created just for us. Changes in what I believe have not been driven by changes in my intelligence and I am well aware that many who I disagree with are much more intelligent than I am. I write these things now to avoid being misunderstood later.

It wasn't that long ago that the generally accepted opinion among most people, including most educated people, was that the mind is independent of the brain and the body. Thinking, it was thought, is something the immaterial mind does. When the immaterial mind reflects on its operations, it is reflecting on its own essential properties and functions. Critical thinking, it was thought, is equivalent to examining the nature and limits of the immaterial mind's operations. Little, if any, thought was given to the fact that whatever the mind is it works as well as it does by deceiving us into seeing things that aren't there and believing things that aren't true. Witness the daily passing across the sky of the Sun. Anyone can see that the Sun moves across the sky. It does so every day, whether we see it or not. This is so obvious to our senses that one wonders why it ever occurred to anyone to look at this obvious fact in any other way. Yet, a 15th century Polish monk, Nicolaus Copernicus, asked what things would look like if the Earth was moving and the Sun was standing still. The answer is not so obvious because thinking such a thought is not a simple reflex of sense perception. Yet, the answer is very simple: things would not look any different to us if the Earth moved around the Sun. But, if it were true that the Earth moves around the Sun, the consequences would be monumental. For that would mean that the most obvious daily occurrence is an illusion. We might forgive the deity for creating a universe where straight sticks appear bent in water. Our brain quickly tells us that the stick does not bend; it just looks like it is bending. But what kind of creator would trick his favorite creatures by making the Sun appear to go around the Earth when it doesn't? Others may explore that line of inquiry. What I want to explore is the fact that for billions of people this daily illusion is not seen as an illusion. It is seen as true. Yet, the facts are that the Earth is hurtling through space in orbit around the Sun while rotating at a high rate of speed on its axis. Earth's circumference at the equator is 24,901.55 miles (40,075.16 km). If you are standing at the equator then you are moving more than 1,000 mph (1,600 km/h) relative to the center of the planet. The planet is also orbiting the Sun with a velocity of about 67,062 mph (107,300 km/h). The illusion of the Earth's motionlessness is due to the relative motion of planets and stars, something which science has explained.

Other illusions have also been explained by science, though many people refuse to accept the explanations, and some illusions remain controversial. Everyone agrees that a rising full Moon looks bigger on the horizon and appears to shrink as it moves upward in an arc until it is overhead. As it rises, the Moon appears to be following a curved path in the sky, a path that resembles an inverted bowl with a flat bottom. Obviously, the Moon does not shrink and grow as it orbits the Earth. The Moon is the same size when it rises as when it is overhead. Some sort of illusion is going on. How can the same object appear so much larger on the horizon than it does when overhead?

Intuition might tell you that the Moon is closer when on the horizon and farther away when overhead. The brain has evolved over millions of years to trick and deceive us on a regular basis. This is for our own good. Without these deceptions the species might have perished long ago. (I'm going to talk metaphorically for a bit. I really do know that the brain does not have a mind of its own.) Things far away appear smaller than things nearby, but the brain knows the real size of that tiny looking tiger in the distance. The apparent difference in size of objects that vary in distance from us informs us of approximately how far away we are from those objects. It might have been advantageous to our survival as a species to be able to tell when a predator was 1,000 meters away and when one was 10 meters away. It would have been of no advantage to think that tigers are tiny when far away and grow menacingly large when they get close to having us for breakfast. But there's a problem with our intuitive explanation of the Moon illusion. The difference in distance of the Moon from a point below the Moon or from a point to the Moon on the horizon isn't enough to account for the difference in appearance of the size of the Moon when rising and when overhead.

Given the evolutionary history of our brain and how we perceive the size of objects relative to their distance from us, there are only two possible explanations for the illusion. Either the brain thinks the Moon is closer than it is when on the horizon or the brain thinks the Moon is larger than it is when it rises. Scientists have ruled out the former explanation on two counts. The Moon isn't always closer when on the horizon and the illusion of the bowl-shaped sky tricks the brain into thinking that the sky at the horizon is much farther away than the sky directly overhead. This means that when the Moon is full and on the horizon the brain, the great deceiver, is itself deceived and presents a giant globe because it thinks the Moon is farther away than it really is. Shouldn't that make it appear smaller? You'd think so, but not in this case or in other cases of what is called the Ponzo illusion. I've had one experience of a perceptual illusion where I saw something as much larger than it is because I thought it was much farther away than it really was. I was at Crissy Field in San Francisco looking across the bay at the coastline of Marin County a couple of miles away when suddenly a giant bird flew into view. It looked like what I imagine a giant pterosaur would look like if it were flying a couple of miles away from me. When my eyes refocused on the jetty about one hundred feet in front of me, instead of on the coastal hills a couple of miles away, I realized it was just an ordinary seagull! I don't think this completely explains the Moon illusion, though, because I've experienced large flying things up close and far away, but only a few astronauts have had the pleasure of seeing the Moon up close. (For more on the Moon illusion, see astronomer Phil Plait's explanation at <tinyurl.com/22qk39z>.)

The greatest illusion of all, perhaps, is the illusion that everything has been designed for a purpose. Clearly, the eye was designed for seeing, the ear for hearing, and so on through the entire litany of all things great and small. Before Charles Darwin and Alfred Russel Wallace in the 19th century asked what things would look like if they had evolved by natural processes, it was not apparent that things would look exactly as they do when one assumed a Great Designer in Heaven creating everything for a purpose.

Figure 1

Most people have no trouble understanding that the same arrangement of dots on a piece of paper can be seen either as a vase or as two faces (**figure 1**), but many people don't understand that a purposeless universe would look exactly the same as one where every hair is counted and every event is planned down to the last detail.

Perhaps the greatest illusion is the illusion of having a soul. Could anyone tell the difference between two people, one of whom has a soul and one of whom doesn't? The logician Raymond Smullyan created an interesting paradox in a short story called "An Unfortunate Dualist." A depressed fellow wants to commit suicide and discovers a magic elixir that will kill his soul but leave his body functioning exactly as before. Smullyan adds another dimension to the idea that there would be no noticeable difference between a person with a soul and one without a soul. A friend of the depressed man sneaks in while the man is sleeping and injects him with the soul-killing elixir. The man wakes up, not knowing he has no soul, goes to the drugstore and buys the elixir. He takes it but doesn't notice anything different. "Doesn't all this suggest," asks Smullyan, "that perhaps there might be something just a *little* wrong with dualism?"

Today, the generally accepted opinion among most neuroscientists is that the mind is the brain in action and the human brain is the result of millions of years of evolution. Critical thinking is something the evolved human animal does. When we reflect on our thinking, we must take into account how the brain evolved. Usually this involves assuming that the modern brain is the result of tens of thousands of years of adaptations to the environments which our human—and even our pre-human—ancestors found themselves in. To think critically is to be aware of the effects of these adaptations on the way we experience the world as we try to make sense out of our experiences. Nature has driven us to think in ways that benefit our chances of survival and reproduction. These ways of thinking may not lead us to care much about the truth. They may, in fact, drive us to prefer the "palpably not true" to the "truth as science finds it." It sure seems that way when one takes in the landscape that is belief in America about the origin of species and the known universe. The methods of science alone, the inventions, the discoveries, the unimaginable growth of understanding about the workings of the human body, disease, and health; the almost daily discovery of something wonderful about the stars, planets, earthquakes, birds, oceans—the list is nearly endless—should make even the most otherworldly heart among us swell with pride and enthusiasm, with joy and exuberance. Yet, as in the darkest of the Dark Ages, we find millions of Americans actively denying the beauty and wonder that science pours out from every pore of its still steaming body born only a few centuries ago. The high priests of religious fundamentalism and literalism would have killed science in its crib and declared that they alone are the protectors of mankind. The dualist tells us that we have souls that live on as immaterial spirits after the body has died and decayed. If the fundamentalist dualist stopped there, we might engage him in some sort of dialogue. But he will not stop there. He must go on to claim that everything science tells us is wrong—unless what science tells us agrees with his interpretation of the Bible.

The religious literalists are not the only barbarians defending the gate who want us to admit that everything science tells us is wrong.. The so-called New Age folks have their own regiment in the Anti-Science Brigade. No scientist anywhere has any evidence of any such law as *the law of attraction*, *the law of similars*, or *the law of infinitesimals*, yet millions of people are absolutely sure of their reality. The so-called law of attraction is a throwback to the magical thinking that characterized the human species in its infancy. We readily excuse our early ancestors for believing that like cures like or that our mental disposition might attract similar external circumstances and events. It is only by selective thinking that one could delude oneself into thinking that if only I think positively then good things will happen to me and my cancer will go away. We might find some comfort in believing that bad things happen to us because we aren't living right. We might even find some science to support our belief. Sometime it is true that bad things happen to us (like lung cancer) because we aren't living right (we shouldn't have smoked cigarettes for all those years). But it is foolishness to think that everything that happens to us happens because of our attitude or frame of mind. Does anyone really believe a baby born with cancer in the eye is born that way because of her attitude? We are asking for trouble and disappointment if we believe that we can make reality change just by willing it. Of course, it is obviously true that very little good will happen to you if you just sit on your hands and make no plans or don't devise ways to make things happen. That's not the law of attraction. That's a self-evident truth.

While we might excuse the magical thinking of those who devote their lives to such fantasies as the law of attraction, I can think of no excuse for those who risk their health and their lives—and the health and lives of their children—by putting their faith in homeopathic "medicine." This "medicine" is nothing but water. Using homeopathic water is on par with using water from some holy well: both are just water and have no inherent medicinal properties. You can no more give water the power to heal by shaking it vigorously after diluting something in it until nothing is left but water than you can by waving your hands over a flask and uttering an incantation in Latin or any other tongue. At least the believer in the healing power of holy water from sacred places rests her belief on *faith*, rather than on some fanciful law that has never been observed by anyone. Samuel Hahnemann (1755-1843), the father of homeopathy, came up with his dilution idea prior to our understanding of atoms and molecules. We might excuse him for his ignorance. There is no excuse for anyone today not to know that when you dilute a substance 100 to 1 and do so 30 times there isn't a single molecule left of the original substance. Hahnemann's other unscientific notion—that like cures like—might be excused as another throwback to the early days of human evolution when the brain had little knowledge in its memory banks and functioned mostly by trial and error. But thousands of years of experience have taught those who care to know that there is no scientific basis for the belief that like cures like. There is no law of similars and to believe so is to believe something that is palpably not true.

We know that many people feel better and get better after taking homeopathic remedies despite their lack of scientific merit. We also know why they feel better or get better after visiting the homeopath (or any of a number of other non-scientific healers). We know that homeopathy, prayer, faith, and dozens of other unscientific practices can and do have salutary effects on many people. Many scientific studies have shown that people who get better after a treatment for some ache, pain, or disorder often do so for reasons other than the medicine or active treatment given. Many of us have given credit for our recovery to a medicine when in fact the problem went away because of spontaneous improvement or we mistook a fluctuation of symptoms for the end of our troubles. Most illnesses don't kill us; they resolve themselves in a week or two. Many studies have shown that sometimes when one group of people is given a placebo (an inactive substance like starch or water), they do just as well as another group given medicine. This is true not just for homeopathy, but for science-based medicine as well. Many kinds of pain, especially back pain, come and go. We are likely to seek some sort of alternative treatment when things are at their worst, which is exactly the time that things will start getting better on their own. Scientists call this *regression to the mean*. We often forget or don't give credit to some additional treatment that may have actually been the cause of our relief. Weirdly, scientists have even found that some people out of politeness or a desire to please their homeopath (or whomever) say they are feeling better when they really aren't feeling better. We also know that patient expectation and healer suggestion play a role in how we feel after a treatment. We know that such things as belief, motivation, and expectation can have the same kinds of physical effects as, say, morphine. We call these effects "placebo effects." It is a near certainty that "battlefield acupuncture" in place of morphine for wounded soldiers works this way. Some scientists think that the effectiveness of Prozac and similar drugs for depression is due almost entirely to placebo effects. Scientists have found that the bodies of dogs, as well as the bodies of humans, can be conditioned to release such chemical substances as endorphins, catecholamines, cortisol, and adrenaline. One reason, therefore, that people report pain relief from both true acupuncture and fake acupuncture may be that both are placebos that stimulate the opioid system. (See Appendix C.)

Some people think that the fact that alternative medicine is placebo medicine means we should be promoting placebo medicine. After all, who cares *why* homeopathy or acupuncture works? What matters is that they *do* work. No, what matters is *why* anything works. If scientists can tease out the various causes of relief that are not due to active medicine or treatments, then maybe they can devise better ways to deliver medicines that we know from scientific studies do work. More important, though, is the fact that we know these alternative treatments are not and cannot be effective for such things as preventing malaria or treating cancer. We also know that placebo opioid effects are not as strong or as long lasting as treatment with a real painkiller such as morphine. While it is true that you are not going to be harmed by homeopathy most of the time, it is not true that the reason you get better from taking a homeopathic medicine is due to some substance having been diluted out of presence and shaken vigorously.

Despite their apparent benefits, another harmful effect of alternative treatments is that patients can be misled by unscientific healers to imagine they are suffering from nonexistent disorders (e.g., allergies and poisoning from "toxins" such as mercury in dental amalgams) and then provided treatments for their imaginary problems. The patients in such cases know they've suffered and know they now feel better. It is only natural that they would conclude that their phony healer is the real thing. Our brains have evolved to make such connections. It is not natural to mistrust our brain, so it is understandable why so many of us make erroneous causal connections. We're hardwired to do so and it takes a lot of reprogramming of our brains to overcome these natural tendencies. This reprogramming is called *education* and *learning*. And the part of education and learning that is most relevant here is called *science.* If thinking scientifically were natural to our species, it would not have taken so long to get where we are and there would not be so many people who are resistant to science.

Yet, even some of those who are not resistant to science have hindered the efforts of scientists to educate the public about the wonders of the world around them. There exists a band of scientists who, while doing science, echo the notion of the religious fundamentalists and New Age gurus that everything science tells us is wrong. Many of these scientists go by the name of *parapsychologist*, an accident of history that has done no favor to the legitimate science of psychology. Many others go by no special name, but they loudly and proudly describe themselves as "alternative." For those who must know more about these characters, Google any of the following: "alternative physics," "alternative history," "alternative archaeology," or "alternative medicine." We can't deny the wildness of the imaginations roaming these alternative fields. Nor can we deny their disdain for logic, their penchant for selective use of evidence, and their admirable ability to get their ideas into digital letters and pictures for consumption by hordes of people craving ever wilder tales that have little scientific merit. These alternative thinkers and their followers seem to consider their rejection of "mainstream" science—i.e., science!—worthy of a badge of courage. They give skepticism a bad name.

The "alternatives" aren't the only ones giving skepticism a bad name. There are two other platoons marching with the Anti-Science Brigade: the *denialists* and the *contrarians*, neither of which should be confused with the lone scientist working away on an idea that conflicts with the consensus of his fellow scientists. For an example of the latter, we need look no further than the 2011 Nobel Prize winner in chemistry, Daniel Shechtman. He claimed to have found a new crystalline chemical structure that seemed to violate the laws of nature. In 1982, Shechtman discovered what are now called "quasicrystals." Such structures were thought to be impossible, but there they were. Shechtman says when he persisted in the idea he was thrown out of his research group for bringing shame on them. The discovery "fundamentally altered how chemists conceive of solid matter," the Royal Swedish Academy of Sciences said in awarding Shechtman the $1.5 million prize.

The denialists and contrarians will never be awarded anything like a Nobel Prize, for reasons that will become obvious. Denialists deny what most scientists think is true. A denialist often gives a long list of twisted facts and brings up things that might be true, while leaving out many facts. For example, denialists have twisted the facts and left out many things to show that cigarette smoking is safe, that evolution is a hoax, that vaccinations are not safe, that 9/11 was a plot by the Bush administration, and that the Apollo Moon landing was a hoax.

Contrarians demand absolute certainty before they will accept something as true. Never mind that hardly anything is absolutely certain, contrarians only worry about this when their personal ox—often this ox is a political animal that looks like an elephant or a donkey—is being gored. A healthy skepticism doesn't require us to reject claims unless they are absolutely certain. Some claims have been proven beyond a reasonable doubt. Scientific skeptics know that most scientists could be wrong about such things as global warming, cigarette smoking, and vitamins. But when the majority of scientists agree, for example, that the evidence shows that cigarette smoking causes lung cancer or that human behaviors are contributing to global warming with potentially devastating consequences, the scientific skeptic doesn't reject the claim simply because there is some possibility that some study in the future will show that they're wrong. Nor does a scientific skeptic agree with those who consider such things "controversial" because they can find scientists who disagree with the consensus.

Philosophy and Science to the Rescue....sort of

It wasn't that long ago that the philosopher Réne Descartes (1596-1650) was certain that critical thinking would lead us to absolute certainty about all the things that matter. David Hume and Immanuel Kant burst that bubble in the eighteenth century. Today, we accept it as fact that the best we can hope for in most of the things that matter is some degree of reasonable probability that what we believe is true. When I speak of "we" I mean those of us who are not part of the Anti-Science Brigade. Mathematics and formal logic may prove some things once and for all, but all the other sciences and most of the thinking we do in everyday life must be tentative. The Anti-Science Brigade will have none of it. Their pernicious deeds are many, but the ones that should anger us the most are those that proclaim it is only fair to teach their stories (which they hold as absolute truths) alongside whatever science is being taught. Where we teach evolution, the Anti-Science Brigade says, we should teach creationism. It would be un-American to do otherwise. Nonsense! We will not teach alchemy alongside chemistry, numerology alongside algebra, nor astrology alongside astronomy. Nor should we teach creationism alongside cosmology or evolutionary biology.

Philosophers from the time of Socrates to the present day have been in the forefront of offering incisive criticisms of what most people instinctively believe. It was not that long ago that many philosophy teachers considered themselves the best equipped profession for teaching critical thinking to the next generation. That notion is no longer sustainable. Along with traditional epistemology, we must recognize that psychology (including social psychology, behavioral economics, and evolutionary psychology) plays a fundamental role in any attempt to guide ourselves or others in critical thinking. Our first guide must be recognition of our biological limitations. Psychologist James Alcock put it this way: "The true critical thinker accepts what few people ever accept—that one cannot routinely trust perceptions and memories." We might formulate this into a guiding principle: *Trust no one, not even yourself.* Of course, if we stop there and take this principle too literally, we will get nowhere. We can't live as social creatures without some good measure of trust in our fellow citizens. Nor could we survive if our senses and memory weren't reliable to a high degree. To think critically, we must examine how we come to believe anything or accept any action as reasonable or right. We must study the pitfalls and hindrances that prevent us from arriving at rationally defensible beliefs and actions. And we must learn to avoid those pitfalls, recognize the hindrances, and find ways to overcome them. This job is painfully difficult and it is made all the more arduous by being surrounded by members of the Anti-Science Brigade.

Overcoming our natural overconfidence in our memories and interpretations of experience goes against the grain. It's unnatural to challenge ourselves about things that seem obviously true to us. But if we want to know the truth about things, rather than just be certain about them, then we will have to practice some unnatural acts in public. Nothing conflicts with our natural inclinations more than critical thinking. Truth attracts us when it brings comfort or security, releases tension, or arouses some pleasurable feeling. But truth is often indifferent to our well-being and is often not as attractive as a comfortable falsehood.

You and I evolved to deceive ourselves and to deceive others. Caring enough about the truth to pursue it does not come naturally to most people. But even those who commit themselves to a lifelong pursuit of fair-minded, reflective inquiry will never succeed at becoming a perfect critical thinker. No matter how open-minded one becomes, you cannot know whether you have overlooked something relevant and important. You can never know whether you're deceiving yourself and basing your decisions or conclusions on desires rather than evidence. You do not become a critical thinker by admitting you might be wrong or declaring that you hold your beliefs tentatively rather than absolutely. Even tentative beliefs can be arrived at uncritically, without fair-minded or reflective thought.

A critical thinker examines all the relevant data, but in any given case you never know whether you have all the relevant data. In some cases, such as politics, you know you don't have access to *all* the relevant data.

How many people want to begin studying something that cannot promise them unqualified success and which can almost guarantee frustration? If you do not become frustrated because of the many obstacles that stand in the way of your success as a critical thinker, you will be frustrated by the lack of interest in critical thinking by many of the people around you. Even worse will be the contact you will inevitably have with many who are demonstrably hostile to critical thinking. Many of them will fancy themselves excellent thinkers, rational to the core, brimming with self-confidence, and absolutely certain that they have the truth and you don't. Most of us take our sensory experiences as unqualified reflections of reality. Or worse, we think that having "faith" in some set of claims gives us a special dispensation to avoid defending our beliefs. Most people have little or no skepticism about their perceptions or memories. Skepticism is a skill you must develop. We are not born—nay, we *couldn't* have been born—mistrusting our senses and our memories. Our survival depends on us not being skeptical, critical thinkers. This fact leads many of us to mistakenly think we cannot be deceived about what's right before our eyes. In the end, many of us will come to identify good thinking with confirming our biases. We will be pleased with ourselves as we find that the more we learn the more we learn we were right all along! Yes, as long as we ignore or demean all the evidence against us, we'll find that the world keeps unfolding along lines that fit nicely with our beliefs and prejudices.

So why would anyone strive to become a critical thinker? You'll be an outcast, perhaps. Few parents will say *I am so proud of you for challenging me!* Those in power aren't likely to encourage you to question their decisions. Most of your teachers won't praise you for what they may see as subverting the status quo and challenging their authority.

You might deduce from what you have just read that I am more cynical and pessimistic than Mencken was. But wait; there's more! I have to confess to what I consider my greatest illusion or delusion, call it what you will. In any case, I have tried for many years to come to terms with the fact that the preponderance of the evidence indicates that there is no such thing as free will. Yet, I can't accept the idea that all our thoughts and actions are determined by causes over which we have no control. I can't accept the idea that nobody is responsible for his thoughts or actions. Our natural inclination and belief is to think we are free and responsible for our actions, except in those obvious cases of brain or neurochemical damage or where we are coerced into doing or saying things. Unfortunately, the world would not be experienced any differently were we not free to choose one path rather than another. The sum total of the evidence from the sciences seems to overwhelmingly support the determinist hypothesis, yet I can't accept it. We've evolved to think of ourselves as free agents, and overall I think this has worked out pretty well for us. In any case, most of us recognize that some people have more constraints than others due to brain damage, psychological trauma, or mental illness. We also recognize that some of us have more constraints than others due to age or mental feebleness. And some have more constraints due to fewer opportunities. But very few among us can take seriously the idea that we should not hold anyone responsible for anything since there's no free will.

In any case, whether you are free or determined, I'm going to present you with arguments that you will either agree with or not find compelling. Whether either of us is free to do what we do seems beside the point and of no consequence to the issues I'll be covering. Either my words will have a causal effect on your thinking or they won't. If they don't it may be because you already agree with me or it may be because you freely reject my arguments. I don't see any way to tell the difference between being causally moved to agree or disagree and freely choosing to agree or disagree. If we can't see a difference, does it really make any difference which is true? I don't see why it would.

Whether we're free or not, there will be consequences from choosing to pursue a life of critical thinking. You are very likely to end up standing out as one against the many if you devote yourself to the pursuit of reasonable beliefs and actions. You will be in a minority, to be sure, but a minority you can take pride in belonging to.

So, with that optimistic little introduction laid out, here is what is in store for you in the following chapters.

2. Critical Thinking: How To Lose Friends and Alienate Your Neighbors

Critical thinking requires that we be open-minded, skeptical, and tentative in our beliefs. Why be open-minded, skeptical, and tentative when we're surrounded by close-minded sheep who know what they know and know that what they know is right? In this chapter, you'll find out why you will be disliked, perhaps even hated, for critical thinking. While society may benefit from having many critical thinkers, individuals who think critically are often marginalized or silenced. Finally, as this is a book about critical thinking, it might be of value to define what we mean by that expression. There is probably no need for it, but I will remind you again of why critical thinking is an unnatural act.

3. Believing is Seeing: Trust No One, Not Even Yourself—Especially if You Find Meaning in a Dirty Diaper

This chapter explains why you can't trust anyone, including yourself. If you start seeing Jesus or Mary in a dirty diaper, it is time to re-examine your trust in your senses and how you interpret experience. We're driven by processes we don't understand but which lead us to find meaning and significance in things that have no meaning or significance. Most of us don't know how to evaluate odds properly, so we think things couldn't possibly be coincidental when they are. We see patterns where there are none and then call in the press to spread the good news.

4. Extraordinary Renditions and Graphic Illusions in a Vaguely Familiar Universe

It is hard to believe but it is true that some members of our species use language to manipulate thought and behavior. (Just kidding, of course. This is *not* hard to believe, unfortunately.) These are the doublespeak experts. They try to persuade us that something bad is really good or that we can be unique and special if we buy the same thing they hope everybody else buys. Advertisements have tried to associate a healthy lifestyle with smoking: light up a Salem after you climb the Alps or a Benson & Hedges after a game of racquetball. The critical thinker must know how to frame the debate so your opponent can't possibly win. The pen really can be mightier than the sword if you know which words to use. If you're clever, you can mislead others with pictures and diagrams just as effectively as with words.

5. Driving an Edsel to the Bay of Pigs

Why is it that groups of highly intelligent people often make bad decisions? In this chapter, we'll discuss the disastrous role of *communal reinforcement* by team players who toe the line. We'll look at what successful companies have in common when it comes to making good group decisions and what groups that make bad decisions have in common. (In case you are wondering about the title of this chapter: Ford Motor Company's decision to produce a metallic monstrosity it named after Edsel Ford and President John F. Kennedy's decision to invade Cuba are often cited as examples of disastrous decisions due to groupthink.)

6. Reliable Sources of Confusion, Collusion, and Spam

Sifting out reliable and useful information from all the sexy garbage thrown our way by the mass media, politicians, scholars, scientists, talk show hosts, bloggers, and so on, is becoming nearly impossible for the average person. How do you tell who's likely to be providing reliable and accurate information? How do we cut through the propaganda, the advertising, the hype and speculation of the mass media, the information and misinformation overloads? Experts on every subject under the Sun seem endless. With so many claims made by so many people, the critical thinker has a problem: Who do you trust? Who can you believe?

7. Seductive Stories and Varieties of Scientific Experience

A good story trumps a dozen scientific studies. Anecdotes can be powerful persuaders but there are several reasons why they are not compelling evidence to a scientist. In this chapter, we'll review the main reasons anecdotes are not good evidence and what we hope to accomplish by doing scientific tests to determine whether the implications of anecdotal evidence pan out under controlled conditions. We'll see how easy it is to deceive ourselves and make mistakes in causal reasoning. We'll describe how scientists try to minimize self-deception by using randomized, double-blind controlled studies when possible.

8. The Fallacy-Driven Life

This chapter will focus on fallacies in reasoning—those errors others are always making! The two main errors in reasoning, in my view, are the selective use of evidence and giving improper weight to evidence. I'll review the popularity of irrelevant appeals, the commonality of slanted, selective, biased, one-sided, incomplete arguments. There are many "truths" in the arsenal of the fallacy-driven crusader: the straw man, the false dilemma, begging the question, the non sequitur, to name just a few.

9. Are We Doomed to Die with Our Biases On?

Most of us believe many things that are probably not true. Why? Is it advertising? Television? Laziness? The power of the media or our parents and teachers to brainwash us? Do our brains naturally lead us astray? Is there something in the way we go about evaluating our experience and accumulating beliefs that misleads us? What hope is there that we can overcome our natural tendencies and become critical thinkers?

10. 59+ Ways to Develop Your Unnatural Talents in Critical Thinking, Skepticism, and Science

In this chapter, you will learn many ways you can expose yourself to an unnatural life and not be ashamed of it.

II
How to Lose Friends and Alienate Your Neighbors

"...intelligence...is in plentiful supply...the scarce commodity is systematic training in critical thinking." —Carl Sagan

"There's a seeker born every minute."—attributed to many people

---

In this chapter we learn what critical thinking is and why it is unnatural to think critically. You might wonder about the profit in being open-minded, tentative, and skeptical when you are surrounded by close-minded, dogmatic bigots. Why not just go along with the crowd? Critical thinking might benefit society, but will it benefit you as an individual? Yes, more than you could ever imagine.

Uncritical Thinking:
Cults, Chiropractors, and Magical Jewelry

We'll begin this chapter with a morality tale about a small band of misguided seekers in quest of living on a higher level than that afforded most of us on planet Earth.

Dorothy Martin led a small UFO cult in the 1950s. She claimed to get messages through automatic writing from extraterrestrials known as The Guardians. Like the Heaven's Gate cult forty years later, Martin and her followers—known as The Seekers or The Brotherhood of the Seven Rays—were waiting to be picked up by a spaceship. In Martin's prophecy, her group of eleven would be saved just before the total destruction of Earth by a massive flood on December 21, 1954. When the day of reckoning came and went, it became evident that there would be no flood and that the Guardians were no-shows. Martin allegedly became euphoric. She claimed that she'd received a telepathic message from the aliens explaining that God had decided to spare the planet as a reward for their great faith. All but two of her merry little band failed to recognize that this new revelation was rationalized rubbish. They not only stuck with her despite the absurd improbability of her claim; their devotion actually increased.

A competent thinker would have judged Martin's prophecies to be wrong when the spaceship didn't arrive and the cataclysm didn't happen. Martin's followers may not have been very competent thinkers before they joined her cult, but devotion to Martin did nothing to increase their competence as thinkers. Their belief that a spaceship would pick them up was based on *faith* in Martin's prophecy, not evidence. Likewise, their decision that the prophecy's failure wouldn't count against their belief was another act of faith. With this kind of irrational thinking, it is pointless to produce evidence to try to persuade people of the error of their ways. Their belief is not based on evidence, but on faith. To some people, such irrational and impenetrable faith is admirable. This book is not for them. It should be noted, however, that there is evidence that even if a person bases his belief on evidence rather than on faith, providing contrary evidence can backfire. Political scientists Brendan Nyhan and Jason Reifler coined the term "backfire effect" to describe how some individuals when confronted with evidence that conflicts with their beliefs come to hold their original position even more strongly.

Irrational faith is not a virtue, but a vice. It is inimical to truth. I don't deny that truth is often hard to come by and that faith offers an easy way out when the road to truth gets rocky and rough. But we do no justice either to ourselves, our society, or our species when we abuse the faculty that sets us apart from the rest of the animals on our planet. Creating inventive rationalizations to support disproved claims or failed prophecies and policies is an abuse of reason, no matter how clever the rationalizations. Yet, as I hope to demonstrate, faith in healing and in supernatural or paranormal causal agents is so *natural* to our species that it, too, might be called a faculty that is unique to our species. We might even go so far as to claim that *having a naturalistic worldview is unnatural*.

Some identify irrational faith with religion, but religion doesn't have the only lock on irrationality. Nor is irrational faith the same as abandoning reasoning altogether. Also, there are some religious thinkers who, when confronted with overwhelming evidence that what their religion has been teaching is wrong, go with the evidence. There are many religious people who accept evolution, even though they were brought up to believe that God created the universe in six days and created all species individually, once and for all. Then there are those who, when confronted with overwhelming evidence that what their religion teaches is wrong, devote their lives to defending its errors. There are many religious people who refuse to accept the evidence that the story of a universal flood as presented in the story of Noah and his ark contradicts every reasonable interpretation of the available evidence. (And this is independent of the fact that such people are not offended at all by understanding this story as literally depicting the creator of the universe as a genocidal maniac.) Instead of admitting that this story could not be an accurate depiction of events and therefore must have some other purpose than to set the geological record straight or to show how tough God is, these apologists for a literal interpretation of the Bible devote their lives to developing hypotheses that confirm what they already "know" to be true. They don't examine evidence with an eye to finding out what the truth is. They look for arguments and evidence that they think will support their beliefs. At times these apologists might be very logical, but logical thinking is not the same as critical thinking. If you think you already know the truth and are looking at evidence and arguments only because you are looking for ways to confirm your biases, you are not thinking critically.

A critical thinker must be willing to engage in an open and fair evaluation of *all* the relevant evidence, not just the items he or she can use to support a personal belief. The goal of the critical thinker is not to defend beliefs against all attacks, but to believe whatever a preponderance of the evidence supports. Apologists for falsehoods and errors are not all religious, however. For example, psychologist Ray Hyman and Wallace Sampson, M.D., tested several chiropractors who claimed they could tell whether a substance was 'good' or 'bad' for you by having you hold the substance in your hand with your arm outstretched while the practitioner of the art of *applied kinesiology* (AK) presses down on your arm and subjectively measures your muscle resistance. As absurd as this practice sounds, there are thousands of people who swear by it. The method has become quite popular among sellers of magical rubber wristbands such as Power Balance and Energy Armor. David Hawkins, a retired psychiatrist, uses the method as a lie detector. He says he can tell whether any statement you make is true or false just by pressing down on your outstretched arm. No one has tested Hawkins or AK as a lie detector under controlled conditions, as far as I know, but Hyman and Sampson did a randomized, double-blind test of AK with the chiropractors. The chiropractors claimed that they could tell whether a vial contained glucose or fructose by pressing down on the outstretched arm that held the vial. (The chiropractors claimed that glucose is a 'bad' sugar and fructose is a 'good' sugar, providing yet another example of bad thinking, but that lesson will have to be put on hold for now. Suffice it here to ask: *do you know what your brain would be like without glucose?*) When the chiropractors knew what was in a vial, they identified its contents with 100% accuracy. When neither the chiropractors nor those testing them knew what was in a vial—a necessary condition for a properly controlled test—their accuracy dropped to what one would expect of a monkey throwing darts at a Ouija board. Did the chiropractors give up their claim? No. They told Hyman that it was obvious that double-blind tests don't work! Indeed, such tests do not work if your goal is to prove what you already know regardless of the outcome of the test.

You are deceiving yourself if you persist in your beliefs after they've been falsified in randomized, double-blind tests. (In Chapter Seven, we explain what such tests are and why they are important in science.) I don't know whether Hyman told the chiropractors that the *power of suggestion* and *ideomotor action* (unconscious muscle motion) could adequately explain why AK seems to work. The AK practitioners unconsciously suggest to themselves that there should or should not be resistance and, lo and behold, their unconscious thought is transmitted into feeling strong or weak resistance. The power of suggestion and ideomotor action also explain the results that scientists get when they test people using a Ouija board, a dowsing rod, or the Power Balance bracelet under uncontrolled and under controlled conditions. We might ask: *why do some people persist in defending claims that have been falsified?* Many think that the answer to that question is simple:

cognitive dissonance. In Chapter Nine I will address this question in detail. Here I will limit my comments to noting that *self-deception* is a powerful hindrance to critical thinking and that nobody is immune to its allure.

The chiropractors and the cult members waiting for their spaceship wouldn't accept the facts because the facts did not agree with what they knew with absolute certainty. They didn't deny the facts. They just interpreted them in such a way so they would not count against their beliefs. A critical thinker can't deny that we have to interpret facts and evidence, but we cannot interpret them according to a rule that says before the investigation begins that *nothing can refute my belief and if it does, I won't accept it*. You are certainly entitled to your own opinion, but not to your own facts.

Defining Critical Thinking

Up to this point I've been using the expression 'critical thinking' as if it were obvious what the term means. It's time now to put forth a definition. For some background on how I arrived at this definition, see Appendix D. *Critical thinking is thinking that is clear, accurate, knowledgeable, reflective, and fair in evaluating the reasonableness of accepting or rejecting a belief or taking an action.* The fields of logic and epistemology have played a large historical role in our understanding of critical thinking. As noted in Chapter One, the social sciences have also played an important part in our understanding of critical thinking. Nothing could be more essential to critical thinking than our perceptions and memories, yet few things are more dangerous to critical thinking than to take perception and memory at face value. We often produce beliefs and make decisions without careful consideration of what is actual or true. We're naturally driven to be selective about the information we draw from experience and we then combine it with information that has been selectively remembered. The result is that we produce new beliefs that are generally consistent with our old beliefs. We might end up with a set of mostly consistent beliefs that are functional, but they are as likely to be false as true. Furthermore, these functional beliefs, even if false, will condition how we interpret future experiences and what data we selectively store in memory. In light of this, recall our first guiding principle: *Trust no one, not even yourself.* We must examine what psychological and logical processes drive us to our beliefs and motivate us to accept actions as reasonable or right. As noted in Chapter One, we must study the pitfalls and hindrances that prevent us from arriving at rationally defensible beliefs and actions. We must learn to avoid those pitfalls and hindrances. More important, we must find ways to overcome them.

One of the first things we must recognize is that critical thinking conflicts with our natural inclinations. We must also recognize that critical thinking must be a cooperative enterprise. We must bounce ideas off others, listen to their criticisms and objections, and form our beliefs according to the best evidence and arguments available. We don't need to feel that we must always arrive at eternal and unchanging certainty. We may often be in a position where what we consider reasonable must be rejected in light of new evidence. If we are to think critically, in other words, we must practice some unnatural acts in public.

Truth attracts us when it brings comfort or security, releases tension, or arouses some pleasurable feeling. But we are not truth-seeking beings by nature. We evolved in ways that make it more natural to deceive ourselves and to deceive others than to seek the truth. Caring enough about the truth to pursue it does not come easily to most people. But even those who commit themselves to a lifelong pursuit of fair-minded, reflective inquiry will never succeed at becoming perfect critical thinkers. We must be open to the possibility that even some of our most cherished beliefs are wrong. That kind of intellectual humility does not come easily. We have to be willing to seek out and carefully consider arguments that go against our beliefs, but we must balance our open-mindedness with a healthy skepticism. We shouldn't accept claims simply because they fit with our beliefs or because the person making the claims seems trustworthy or authoritative.

On the other hand, it is justifiable to hold that some things are absolute truths, even though the word 'absolute' adds nothing to the content of what we believe. It is true, even absolutely true, that you cannot jump off the ground and begin flying like a bird. But, you might object, what about those yogic flyers who say they have learned to fly a few inches off the ground for a brief moment or two at their Transcendental Meditation® classes? They say they are flying. Who are you to say they're not? I'm a fellow with eyes who can see that they are hopping around with their legs crossed in the lotus position. Hopping around is not flying. The world of critical thinking is not the world of Humpty Dumpty in *Through the Looking Glass* where words can mean whatever you want them to mean and mean something different when you are in the mood to have them mean something different. Even a Ph.D. who has achieved the highest level of enlightenment in his sacred group cannot turn hopping into flying just by saying it is so. You cannot be a critical thinker if you insist that words mean what you say they mean no matter what everybody else means by them.

On a brighter note, there is general agreement even among chiropractors and Yogic flyers that a critical thinker must examine *all* the relevant data on an issue. Unfortunately, we never know whether we have all the relevant data. In other cases, we might be able to get most or all of the relevant data, but the issue is too complex for the average person to understand. For example, you might be asked to vote on whether to construct a nuclear power plant in your area. In some cases, we might be able to understand the issues but we don't have the time or resources to do a proper investigation. Or, we might be asked to switch power companies from a privately owned enterprise to a public utility. How should a critical thinker respond in such situations? Some might argue that if you don't understand an issue, you shouldn't vote on it. That sounds reasonable, but it might mean that most people should stop voting on issues altogether. We would have to consider whether that is a consequence we are willing to accept in a democracy. Others might advocate finding reliable sources: individuals or groups who are independent and *can* do the analysis and understand the issues. We should follow their recommendations. That sounds reasonable, but what should we do if different independent groups come up with contradictory recommendations? In that case, we have to study their arguments and determine which one makes the better case. One of the most important elements of critical thinking is the ability to evaluate arguments. So, even if one can't understand all the details of an issue one is asked to vote on, one can still apply critical thinking skills to arguments and arrive at a decision that may be reasonable under the circumstances. Unfortunately, what many people do in such cases is not very rational: they look for guidance from television ads, lawn signs, bumper stickers, lapel buttons, or mailers from interested parties. Or worse, they follow the advice of their bartender, girlfriend, hairdresser, or favorite evangelist.

We might say that thinking critically is like handicapping horses: you never have enough information. You sometimes have to rely on others who have more information than you do, even though their information is incomplete. You can't be sure that you bet on a winner until after the race has been run, and by then it is usually too late (unless, of course, you did pick a winner). Different decisions have different odds. None may be perfect. If you don't have to bet, it doesn't matter what the odds are. But if you don't bet, others will and they will be the ones who determine whether you get that nuclear plant or not, or whether you are forced to change power companies, like it or not.

As noted in Chapter One, striving to become a critical thinker will be a frustrating enterprise. Our own biological makeup will stand in our way, as will the lack of interest in critical thinking by many of the people around us. This will be especially frustrating because critical thinking must be a cooperative venture. The goal of critical thinking necessitates that we not think of those who disagree with us as opponents or enemies, but as co-investigators. Many of those who disagree with us will have little or no skepticism about their perceptions or memories. Many will have no curiosity about how perception and memory work. In fact, many people aren't curious about very much at all. They're quite comfortable with the beliefs they've been brought up with and it makes them uneasy to have their beliefs challenged. Such people tend to be dogmatic and authoritarian. Unfortunately, they also seem to make up a good portion of the population.

Critical thinking requires a disposition to favor empirical evidence, facts, and rational arguments over feelings, hunches, intuitions, and emotions. Many people not only do not appreciate the person who is disposed toward rational thought, they consider rational thinking *inferior* to emotions and intuitions. The critical thinker doesn't ignore hunches and intuitions because often they are the result of unconscious processes that are based on knowledge, experience, practice, training, and facts. But some people have little interest in the facts unless they support their beliefs. They make huge decisions based on gut feelings and have little time for discussion and rational argument.

Many don't think they can be deceived about what's right before their eyes. Furthermore, they identify good thinking with confirming their biases. Hence, many will not think highly of those who criticize them or question their reasoning. Most people do not see critics as co-investigators, probably because most critics are more interested in winning you over to their side than they are in considering your objections and possibly having to admit that they were wrong. There is a real art to criticizing and questioning others without making them defensive, but it is an art the critical thinker must work at mastering. It is not natural for us to seek out people who disagree with us, but critical thinkers need opposition if we are to think at our best. We should be grateful for those who disagree with us. They offer us a chance to reflect on our beliefs and to correct them or dig deeper to defend them.

So why would anyone strive to be a critical thinker? You may alienate your friends and co-workers. Few parents will thank you for challenging them. Employers will probably not reward you. They may take your questioning and reflective attitude as evidence of rocking the boat and not being a team player. Governments will not encourage you. Those in power don't want you asking questions or *thinking* about what they are doing in your name. Teachers may not praise you for asking questions. Some may even belittle you or try to humiliate you for asking questions they can't answer or don't want to think about.

But if you don't mind standing out as one against the many, then you might devote yourself to the pursuit of reasonable beliefs and actions. You'll gain self-confidence and a sense of control over your life. This book is meant to be a guide in that process. It will point you in the right direction. Where you arrive will be largely up to you. What a critical thinker hopes for is to become free from the tyranny of those who would rather see obedient servants than thoughtful, independent thinkers. We should hope also to become free from our own tyranny: the tyranny of self-deception and wishful thinking. We will still make mistakes, but they will be *our* mistakes.

Finally, a critical thinker is an independent thinker. You will be lonely but you will not be alone. You *will* influence and attract others. The influence of the community of critical thinkers might not keep the barbarians from knocking down the gates and occupying many of the rooms in the palace, but think how much worse it would be if we did nothing. Those who do not value reflective thinking, reasoned argument, or basic principles of logic will take over the whole countryside if we just stand back, act polite and complacent, and offer no opposition to their irrational beliefs. We could all end up on a hilltop, hopping madly as we try to escape Earth's gravitational pull or pressing on each other's outstretched arms while waiting for a space ship to take us to the Promised Land in a galaxy far, far away.

SOURCES FOR CHAPTER TWO 269

III
Believing is Seeing
(Trust No One, Not Even Yourself—Especially If You Find Meaning in a Dirty Diaper)

"Convictions are more dangerous enemies of truth than lies."
–Friedrich Nietzsche

This chapter will explain why you can't trust anyone, including yourself. Perceptual biases lead us to see things that aren't there and to miss things that are right in front of us. The selectivity of perception and memory often distorts and biases our understanding of what we perceive and remember. Cognitive and affective biases hinder us from interpreting and evaluating our perceptions and memories accurately. Self-deception, wishful thinking, and operant conditioning drive us to beliefs independent of their truth.

Eyewitness testimony is compelling but unreliable. Unconscious perceptions lead us to incorrect inferences about experience and to find meaning and significance in things that are random or coincidental. We see patterns where there are none, but don't see real patterns because we don't know how to look for them. Most of us are innumerate and think things couldn't possibly be coincidental when that is probably just what they are. We fear things we shouldn't and don't fear things we should because we don't know how to calculate simple odds.

The Power of Suggestion

George A. Dillman teaches martial arts and is a 9th degree black belt in Ryukyu Kempo Tomari-te. He has trained heavyweight champion Muhammad Ali and movie action hero Bruce Lee, among others. He claims he can knock out anyone without touching them by hurling *qi* (ch'i, pronounced *chee*) at them. *Qi* is an alleged energy that permeates everything. The idea originated in China a few thousand years ago and is still popular there as well as among many Westerners today. We can only guess how the ancient Chinese came up with this idea, but the notion of a breath-like energy flowing through all things also evolved in other pre-scientific cultures such as ancient India, where it is called *prana*. The idea has even found adherents in scientific cultures where it is called *élan vital* or the *life force*. Many students are attracted to Dillman's martial arts school because of his reputation for possessing the amazing ability of delivering a no-touch knockout. The students come, they see, and they end up believing because they have seen with their own eyes a demonstration of his superpower. On a National Geographic television program ("Is It Real? Superpowers"), I saw Dillman push his hands through the air while grunting loudly at one of his students. After a few seconds, the student fell to the ground and appeared to be unconscious. Dillman repeated the demonstration and knocked over a line of students without touching anyone. He rolled his hands around an invisible ball in the air before him and pushed his hands forward while grunting. Soon the line of students tumbled over. Dillman bragged that he once moved the line at Starbucks by using only his mind to push *auras* around. (Auras are alleged energy fields that surround bodies.) Several young martial arts students were interviewed on the National Geographic program. Each told how they had come to the school because they had heard of Dillman's power and wanted to see it for themselves. Now that they had seen with their own eyes what Dillman can do, they were convinced he was knocking people out with *qi*. If you can't believe your own eyes, what can you believe?

Luigi Garlaschelli, however, had his doubts. Garlaschelli was not one of Dillman's fawning students, but a chemist at Pavia University in Italy. Garlaschelli and fellow skeptic Massimo Polidoro have had extensive practice uncovering errors made by people who do not consider that there might be a better explanation for their experiences. What appears supernatural or paranormal often has a simple physical or psychological explanation. Garlaschelli was shown a film of Dillman delivering his no-touch knockout. "I don't think it is possible for a person to transfer his or her energy to another person," said Garlaschelli. Why would he? There is absolutely no scientific basis for such a claim. But Dillman has demonstrated it, so even if what he does violates the laws of physics regarding energy transfer, the fact is that Dillman and his followers believe he is knocking people about by manipulating *qi*.

Garlaschelli agreed to put Dillman's claim to the test. For some reason, Dillman did not participate in the test. Leon Jay, one of Dillman's top aides and an 8th degree black belt, was given the job of knocking out Garlaschelli without touching him. Jay moved his hands in the air around Garlaschelli's head for some time but was unable to move the 125-pound chemist even a fraction of an inch.

Dillman did participate in the rationalization session that followed Jay's failure. He claimed that the skeptic is "a total non-believer," which was stating the obvious and, ironically, is the key to understanding what is going on with Dillman's *qi* punch. Dillman was suggesting that one must believe in the power of *qi* in order to be knocked out by it. That is exactly what the skeptics were to maintain. Like the AK chiropractors we met in Chapter Two, however, Dillman was not going to let a controlled experiment destroy his belief. He went on to claim that maybe Garlaschelli wasn't knocked out because his tongue was "in the wrong position." He also claimed that he could not knock out someone if he had one big toe pointing upward and the other pointing downward. He didn't elaborate on the requirements of tongue location or toe direction for the flow of *qi*. Understandably, National Geographic didn't do any further tests on these *ad hoc hypotheses*, so we don't know for sure whether they were just rationalizations, which is what they appear to be, or legitimate exceptions to whatever rule governs the manipulation of *qi*. (An *ad hoc hypothesis* is one created to explain away facts that seem to refute one's belief or theory.) A reasonable person might be willing to consider Dillman's excuses had he revealed them to an experimenter *prior* to the test. Bringing them up after the test had been completed only adds to their implausibility.

Polidoro offered a different explanation for Dillman's no-touch knockout and for Jay's failure to put down a featherweight Italian chemist using this superpower. Polidoro suspects that Dillman's students were being manipulated by *the power of suggestion*, not by the power of *qi*. It was their *belief* in Dillman's powers that led them to unconsciously suggest to themselves that they were being overpowered by some force controlled by Dillman. The students weren't pretending to be knocked out, but they were conditioned to behave the way they did based on what they believed was expected of them in that situation. The expected reaction to the master's moves had been reinforced by the community of students in past demonstrations.

That people can unconsciously direct their muscles to move has been demonstrated many times under controlled conditions. In 1853, physicist Michael Faraday did a rather clever experiment on table tilting, then a common occurrence at séances. He suspected that the people sitting around the table, rather than spirits, were responsible for the tilting of the tables. The sitters denied it and felt certain that they were not moving the table. Faraday used overlapping pieces of paper in front of each participant to help him show that they were moving the table. When they placed their hands on the papers, any movement of their hands would also move the papers. If the papers slid over each other, it could only be because they had been pushed there by the participants. At the end of the experiment, all the papers had moved but none of the participants claimed to be aware of any movement. Faraday concluded that the phenomenon was due to "self-deception resulting from unconscious motor movements guided by expectation." The term *ideomotor action* was coined by William B. Carpenter in 1852 to explain the unconscious movements of dowsers affecting their rods and pendulums. The expression is now used to describe many different kinds of unconscious movements that humans and other animals make. (A very detailed account of Faraday's table-tilting experiments can be found in Richard Wiseman's *Paranormality: Why We See What Isn't There*.)

A rather dramatic example of ideomotor action was discovered by Oskar Pfungst in 1904 during an investigation of a horse that appeared to be able to understand human language and do simple math problems. Pfungst discovered that the horse, known as Clever Hans, was responding to slight, unconscious movements of his master. (Psychologists now sometimes refer to an involuntary physical response to an unconscious environmental cue as the *Clever Hans effect*.) Involuntary physical cues signaled the horse to start or stop tapping his hoof. A couple of decades later, the famous parapsychologist Joseph Banks Rhine investigated an allegedly telepathic horse named Lady Wonder who answered questions by knocking over alphabet blocks. Rhine, a trained botanist, was apparently unaware of ideomotor action or the ability of horses to pick up signals from body language. He declared the horse telepathic. When he returned to test the horse again using better controls, Rhine declared that the horse had lost its telepathic ability. Rhine, like Dillman's students, saw what he expected to see. Unlike the horses, however, the martial arts students who were "knocked out" weren't responding to unconscious signals but to complex beliefs that directed them to fall down or faint without being consciously aware that they were responsible for their own physical movements.

If Dillman's students who were there to see for themselves a demonstration of the no-touch knockout thought critically about what they had observed, they should have recognized that they saw their hero stand in a certain way and make some movements and some sounds. Shortly afterward, they saw another person fall back and drop to the ground. They were *told* that what they were observing was happening because the master knew how to control a special energy that he could turn into a weapon against another person. They saw what they believed. Don't we all? Dillman's students are apparently ignorant of the power of suggestion and of ideomotor action. Garlaschelli and Polidoro, on the other hand, have studied both and were therefore able to consider an alternative hypothesis to the one offered by Dillman for what was happening in his martial arts studio. The moral of the story is that the more knowledge you have, the better equipped you will be to consider alternative explanations for what you experience. The better equipped you are to consider alternative explanations for experiences, the stronger will be your critical thinking ability and the less likely you will be to misinterpret what you see with your own eyes.

Now, let's return to the issue of perceiving what we believe. We all tend to perceive what we believe, but some of us recognize that this is not the best way to arrive at the truth. If our beliefs are driving us to perceive things that aren't there, we need to rein in our confidence in perception. Granted, sense perception usually serves us well, but it's not infallible. For example, many people have seen the Virgin Mary or Jesus appear in such things as water stains, burn marks, tree bark, oyster shells, and dirty diapers. Many religious people seem to see the objects in the world as just so many spiritual Rorschach ink blots waiting for them to ferret out the hidden simulacra. Many people, however, see nothing interesting until *after* someone else has pointed out the face in the cinnamon bun or on the water-stained wall. The same thing can happen with sounds. A person may listen to a recording and hear nothing but garbled noises until another tells them to listen for "my sweet Satan" or "Thatcher is a tart." Suddenly, the distorted sounds become clear as a bell. (Here's a quick quiz: what Bob Dylan song begins with the words "Throw my chicken out the window, throw my fruitcakes out there too"? Hint: *tonight I'll be staying here with you.*)

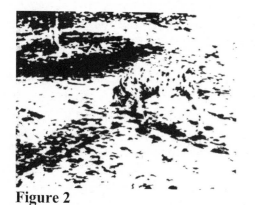

Figure 2

A critical thinker must learn to resist the temptation to see the world according to the suggestions of others without being so obstinate as to be unwilling to look harder at something to perceive what another person insists is there. Sometimes others see things that we miss and we should be open to their suggestions to look closer. Not only do we often perceive things that aren't there, we often *don't* perceive things that are right before our eyes. This is partly due to the fact that we don't see some things because we are focused on something else. Perhaps we're concentrating on something or we're misdirected by a magician. Our not perceiving things that are present is partly due to the way the brain works. The brain seems to take a 'snapshot' of the present situation and unless there is a good reason to alter that snapshot, sounds and sights can change without our noticing them. The brain is not a digital video/audio recorder, replacing its input with updates every millisecond. The healthy brain filters out most of the potential sensory input available to it. This filtering is essential to perceiving anything clearly, but it means that for everything we see or hear there is a lot more that we could have seen or heard but didn't. (Did you see the Dalmatian in **Figure 2**?)

Sometimes we see or hear things without being conscious of seeing or hearing them. Evidence of unconscious perception may become clear at a later time. For example, a person may go many years without understanding why seeing a road sign with the words "green valley" in it produces sexual arousal. Then, one day she returns to a place she hadn't been in many years. She remembers that this was where she met her first lover and the place is called Green Valley. Psychologists call this process of later experiences being influenced unconsciously by earlier experiences *priming*.

We can't deny that millions of people have seen Jesus, Mary, various saints, ghosts, aliens in and out of their spacecraft, Santa Claus, fairies, leprechauns, and faces on the Moon and on Mars. But just because millions of people see something does not mean it exists. We can accept some perceptions at face value. The dog that's biting my ankle is a dog that's biting my ankle. But taking all perception at face value is a mistake. How should we decide which perceptions to trust and which to question? There is no surefire answer to that question. But the more knowledge we have about perception, how it works, how beliefs affect what we perceive, and how we frequently deceive ourselves, the better our chances are of recognizing those situations where we should not assume that what we see is what there is. We might ask ourselves: *Am I seeing this because I want to see this? Has this perception been influenced by suggestions from others? Is it possible I missed something here because I was paying attention to something else?*

The adulation of Dillman's martial arts students is understandable. They want to believe that what Dillman claims is true because if it is then maybe they too can develop this superpower. They see the power demonstrated. They have their belief reinforced by the other students and teachers. But once the belief has been challenged and they see that a non-believer cannot be knocked out with a *qi* punch, they should reassess what they have seen. They must weigh the alternative explanations offered by Dillman and Polidoro for Jay's failure to knock out Garlaschelli. Which explanation is more plausible? Did the skeptic have his tongue in the wrong position or his toes pointed up and down, and was this why he was invincible? Does *not* believing in either *qi* or the ability to transfer it create a shield of some sort that protects the non-believer? Or, did the skeptic not move because the *qi* punch is an illusion? Jay couldn't move the skeptic because Jay wasn't controlling any energy or force at all. What was observed when the students were knocked out without being touched was a result of the power of belief, rather than the power of *qi*. A rational person must side with the skeptic. Had more tests been done with skeptics and believers, some with their tongues aimed this way or that and others with their toes pointing whatever way you like, would Dillman have given up his belief if the results did not support his claim? Or would he have come up with another set of rationalizations for the failure? If I had to bet, I'd put my money on the latter. The students who think critically about what they have observed might learn a valuable lesson about the power of belief, operant conditioning, and communal reinforcement. If their goal is to exercise power over others without touching them, Garlaschelli and Polidoro have provided them with lessons at least as valuable as those provided by master Dillman.

Priming the Primate Brain

Perception is always selective, but it isn't always selecting as if it were concerned with the truth, the whole truth, and nothing but the truth. If we are not looking or listening for something, we often will not see or hear what occurs right before our eyes. People shown pairs of cards with pictures of faces on them and asked to choose the most attractive face often don't even notice when the cards are swapped and they are asked to explain why they chose the card they didn't actually choose. Not noticing the change of cards—or, as often happens in films, changes in attire or setting in a scene due to sloppy editing—is known as *change blindness*. (Even more wondrous is that people will make up all kinds of interesting reasons for choosing the face they didn't actually choose! For more on change blindness and the confabulations people construct to account for the choices they think they made but didn't, see <www.lucs.lu.se/2010/08/choice-blindness/>. If you are not expecting to be tricked—as the folks in the face card experiment weren't—you may not notice something rather dramatic happening in your presence. Those who have seen the "basketball" video produced by the Visual Cognition Lab at the University of Illinois know what I mean. If you haven't seen this video, I recommend that you go to the website at *<viscog.beckman.uiuc.edu/media/ig.html>* and follow the instructions before continuing your reading here.

For those who don't mind missing out on the experience of not seeing a person in a gorilla suit walk across your field of vision without you even noticing, here is what happens. You are asked to watch a video, following only these instructions: *When viewing the video, try to count the total number of times that the people wearing white pass the basketball. Do not count the passes made by the people wearing black. When you're done, visit the lab web site for more information.* When asked if they saw the gorilla in the film, about half of the viewers respond with laughter and disbelief. Some, when shown the film again and told to look for the gorilla, are convinced that they are being shown a different film. The psychologists call this *selective perception* or *inattentional blindness*. The rest of us might call it unbelievable if we hadn't experienced it! (We'll discuss the gorilla film in more detail in Chapter Nine.)

Another way in which perception functions according to its own rules is when we perceive things unconsciously and become aware of our perception only at a later time. Psychologist James Alcock tells the story of standing in line to go into a movie theater when he wondered aloud to his wife about the whereabouts of someone they hadn't seen in years. He wondered why he thought of this person at that moment and then picked his very distinctive voice out in the crowd. He had heard his old friend talking but had not been paying attention to the voice, yet it was registering in his mind. Alcock also tells the story of driving down the street with a friend he had not seen in a while. Alcock says that he was just about to mention the name of a fellow that they had gone to school with a decade earlier, when his friend said 'I wonder whatever happened to W. S. H.?' This was the very fellow Alcock had been thinking of. He turned the car around and began looking for some clue as to why they might both think of the same fellow at the same time. They came upon a large pendulum clock in a store window and stopped their search. W. S. H. was known for walking around campus with a pendulum-style pedometer swinging from his belt. They hadn't consciously noticed the clock as they drove by it earlier, but it seems likely that it had been unconsciously perceived by each of them and triggered memories of their classmate.

It is often said that we perceive what we want to perceive or what we expect to perceive. *Expectation bias* seems to explain why Rhine didn't see Lady Wonder in a different light or why Dillman's students who came to watch the demonstration saw him knock out students with *qi*. Expectation bias, however, happens to the best of us. Our college baseball coach was recognized as one of the best in the business. I happened to be present at a game when a runner on our team was hit by a batted ball as he ran from third base to home plate *inside the baseline*. The rule is that when that happens the runner is out. The umpire made the call. Our coach flew out of the dugout and questioned the umpire's call. The umpire replied in a voice we could all hear in the small stadium: "Joey, you know the rule. He [the runner] was inside the baseline." It was true. I saw it with my own eyes. The runner was clearly inside the baseline and the umpire made the correct call. Our coach replied something to the effect of "That's impossible. My players are taught to run *outside* the baseline." He was a great coach and I have no doubt that his players were taught to run outside the baseline so that they would not be out if a batted ball struck them. But players don't always do what we teach them to do and they don't always do what we expect them to do. And we sometimes see what we expect to see.

It's not likely that the player who was hit by the batted ball didn't remember what he'd been taught. He behaved as if he knew he was out the instant the ball touched him. Memory, however, is every bit as tricky as perception. Memory, like perception, is a selective and constructive process. Memories are also susceptible to suggestions from others. We remember things vividly even though they never happened. Memories often join together events that happened at different times or in different places as if they had happened at one time in a single place. Sometimes, when we need to, we cannot remember important information. False memories can be reinforced by sharing them and by having others validate them. These facts can have important consequences for various professions that require interviewing people and asking them what they remember about an event. Journalists, historians, and criminal investigators—to name just a few—should be well trained in how memory works, especially when dealing with eyewitness testimony. Eyewitnesses can provide valuable information, but their memories can be manipulated. A criminal investigator, for example, must be careful when interviewing an alleged crime victim or a witness to a crime.

Knowledge of how memory works should affect how forensic and psychological interviewers go about questioning people. Investigators shouldn't show pictures or drawings of a suspect to a witness in such a way that might provide the witness with data to fill in the blanks of memory. Telling an eyewitness that a particular individual has confessed to a crime can taint the memory and therefore the testimony of the eyewitness. An interrogation is fatally flawed if the investigator suggests to the witness that an individual in a real or a photo lineup is especially worth a closer look. Otherwise, what is remembered and testified to might be the result of information provided *after* the crime and be wholly or partially inaccurate.

An example of how not to question a person who claims she was assaulted was unintentionally provided by the Durham, North Carolina, police department. A 27-year-old African-American woman named Crystal Gail Mangum told the Durham police that she had been raped by three men on March 13, 2006, while at a party attended by a large number of lacrosse players from Duke University. Mangum and another "exotic dancer" worked for the Allure Escort Service and were hired for several hundred dollars by one of the lacrosse players to perform at the party as strippers. A photo lineup was prepared by the Durham police. It consisted of an array of 46 slides of photos. The officer conducting the interview told the alleged victim that he used photos only of men who police believed to be at the party, 46 in all. This was a blunder no experienced officer should make. Witnesses viewing photo arrays and lineups should be instructed that the real perpetrator may or may not be present and that the administrator of the lineup does not know which person, if any, is a suspect. There should have been photos of men who were not at the party in the mix with those who were at the party and the alleged victim should have been aware of that fact. Why? Because eyewitnesses may feel pressure to identify someone from a lineup—even an innocent person—because they believe the police would not be presenting photos of all innocent persons. Using a mix of suspects and non-suspects will reduce the chance of a mistaken identification. Furthermore, the one administering the lineup should not know who, if anyone, in the lineup is a suspect. Such knowledge may influence how he or she conducts the interview and may lead to inadvertent suggestion to the alleged victim as to who is the suspect. The alleged victim identified three men from the slides shown to her by the officer, but the flawed method of doing the photo array interview casts grave doubt on the reliability of the accuser's identifications. (There are other reasons for doubting this accuser's accusations but they are not pertinent to the issue of eyewitness testimony so we won't go over them here.)

Even an innocent comment by an investigator to suggest that a witness "got it right" by identifying a suspect in a lineup can affect the memory of the witness. Research strongly indicates that confabulated answers to suggestive questions that are rewarded with positive feedback often become fixed memories in those interrogated. Positive feedback can come from verbal praise or subtle body language. It is tempting to want to help a witness remember, but there is a not-so-fine line between helping and suggesting. This line should be large and clear regarding some practices, like hypnosis for example. Some interrogators have been tempted to use hypnosis to help people remember, but the evidence is overwhelming that hypnosis leads people to remember things that never happened. People are often very confident in their erroneous hypnotic memories, probably because they are so vivid and easy to come by under hypnosis. The evidence, however, does not support the notion that the more confident one is in the accuracy of one's memory the more accurate the memory is likely to be. People are as confident of their false and inaccurate memories as they are of their accurate ones.

How perception and memory can malfunction is graphically illustrated by the case of Dr. Donald Thompson, who immigrated to Canada from Australia to study with memory expert Endel Tulving. Authorities contacted him because he matched the description of a rapist that had been provided by the victim. Thompson had been doing a live interview for a television program just before the rape occurred. The victim had seen the program. It seems she mixed up her memory of what she'd seen on television with her memory of the rape. Such misattribution happens occasionally, but the ending is not always pleasant for the one who is mistakenly identified as a criminal. For example, seven eyewitnesses identified Bernard Pagano, a Catholic priest, as having robbed them at gunpoint in Delaware and Pennsylvania in 1979. Robert Clouser, who bore a strong resemblance to the priest, confessed to the crimes. ("The Gentleman Bandit" was a 1981 TV-movie sympathetic to Pagano's plight.) One wonders if some of the victims had had contact with Pagano before identifying him as a robber or if this was just a case of one man looking like another. A police artist had done a sketch of the robber based on eyewitness descriptions. A woman identifying herself as Pagano's lover contacted the police after seeing the sketch and provided photos of him.

One wonders what would have happened to Dr. Thompson had he not had an excellent alibi. If Clouser had not come forth, Pagano may well have gone to prison. Jurors might well have been unsympathetic to a priest with a girlfriend and an attitude. (Pagano played racquetball during breaks in his trial and was described in newspapers as having a cavalier attitude toward the proceedings against him.)

Eyewitness testimony is notoriously unreliable, yet many people have great faith in it and would consider one unduly obstinate for doubting the reliability of the word of several confident eyewitness. Consider Adolph Beck, who served seven years in prison after being mistakenly identified by twenty or so eyewitnesses. He was pardoned in 1904 and awarded £5,000 (about £300,000 in today's money) for his troubles. He died five years later, having seen the establishment in 1907 of the English Court of Criminal Appeal, too late to do him any good. However, even though appeals courts make it more difficult to convict innocent people on the basis of faulty eyewitness testimony, they do not make it impossible. There have been many studies by social scientists that have found that eyewitness testimony is highly inaccurate but is considered highly trustworthy. Most people are not very good at identifying a stranger whom they saw briefly on a single occasion under stressful conditions, especially if the stranger is of a different race than the witness. As with memory studies, eyewitness studies have not found that the *confidence* of eyewitnesses is any indication of the *accuracy* of their testimony. Studies have found, however, that people tend to believe the eyewitness testimony of people who provide vivid details and exhibit great confidence in their memory. The Innocence Project, a national organization devoted to overturning wrongful convictions, reports that of 239 convictions that were overturned based on DNA evidence 73 percent had been based on eyewitness testimony. Many of these overturned cases involved convictions that had been based on testimony from two or more eyewitnesses.

Perception, Deception, and Misdirection

Another thing that scientific studies have shown beyond any reasonable doubt is that the brain has a mind of its own. As noted above, our brains are primarily driven to process data to preserve us and make us feel comfortable, secure, and pleasant, and only secondarily to find the truth. Illusions and delusions are often just as functional as the truth is in serving our purposes. Furthermore, illusions and delusions can be created just by stimulating parts of the brain with electrical current. Such artificially evoked feelings and perceptions are indistinguishable *as feelings and perceptions* from naturally caused feelings and perceptions. Researchers have caused hallucinations, out-of-body experiences, the feeling of a "presence" in the room, the déjà vu experience, and mystical experiences, among others, just by sending electrical currents to various parts of the brain. (That some people subjected to the same kind of brain stimulation *don't* experience any weird feelings indicates that at least some people are exhibiting *expectation bias* and reacting to suggestion and expectation rather than to electrical stimulation of the brain.) Defects in specific parts of the brain can also lead to specific perceptual problems that run the gamut from not feeling a connection between your own body and your "self" to seeing loved ones as decoys in duplicate bodies. Some people who are blind or paralyzed deny it and claim not to be conscious of their disabilities. These studies are interesting in themselves but even stranger than these experiences are the stories people make up to explain them. They often concoct elaborate confabulations to make sense of their situations. Some researchers think that confabulation—mixing actual events with fictions or constructing wholly fabricated stories on the fly without intending to deceive—may be something we all do. There is strong evidence that most children need very little encouragement to make up stories about things they know nothing about. Adults, it seems, aren't much different. It is not just the sick brain that confabulates. We all do it from time to time, especially if we're encouraged to explain things we really know nothing about. For example, many of those in the face-card experiment mentioned above were duped into explaining why they chose the face that they didn't actually choose. A magician had switched cards on them. Nevertheless, they were willing to give elaborate reasons why they chose one face over the other, even though they had actually chosen the other face. Most of us are also affected by *hindsight bias*: we construct our memories to fit with what we believe or know in the present. Many studies have shown that how people had judged something in the past changes in light of new information or later experience and they deceive themselves into thinking that their original judgment was in tune with the new information even though it wasn't. Memory might well be described as the incessant construction of the past and be seen as just one aspect of our tendency to confabulate.

Even though most of us make up stories that are fictions which we try to pass off as factual, society would collapse if there were not a great deal of trust among its members. Usually, that trust is justified. People don't intentionally misperceive or misremember things. And people are accurate often enough to make them generally reliable when they provide information about themselves, other people, or the world. Even so, we know such testimony is not infallible and that it is prone to distortion and error for a variety of reasons. For example, self-deception is so pervasive that we should take most self-assessments with a grain of salt. Why do 85 percent of medical students think it's improper for politicians to accept gifts from lobbyists, but only 46 percent think it's improper for physicians to accept gifts from drug companies? Study after study has found that the vast majority of people think they are less biased, more congenial, less susceptible to improper influence, or more competent than the majority of their peers. Ninety-four percent of university professors think they are better at their jobs than their colleagues. Seventy percent of college students think they are above average in leadership ability. Only two percent think they are below average. Even experts make biased evaluations of ambiguous or inconsistent data. We all tend to be uncritical of data that support our beliefs and very critical of unsupportive data. These tendencies are known as *confirmation bias* and they are pervasive, despite being well known and well understood. We recognize bias in others much more easily than in ourselves.

One area of research demonstrates this and a related cognitive flaw quite convincingly. Numerous studies involving personality assessments, astrological readings, biorhythm charts, and validation of messages from the dead that have come through alleged mediums have found that almost regardless of what data you present subjects with, they rate the accuracy at about 80%. Give people who have taken a personality assessment test the same results to evaluate and even if those results were taken from a newsstand astrology book, the average rating will claim about 80% accuracy for the assessment. Give each member of a group of people the same made-up astrological chart and the average rating will claim about 80% accuracy. Give an elderly woman who wants to make contact with her dead husband a list of items that a psychic claims to have received from the other side in another reading of another elderly woman who also wants to make contact with her dead husband and she will rate the accuracy of the reading at about 80%. Why is this so? There are several reasons. Many of the claims are general and would apply to almost anybody. E.g., *You sometimes feel as if people don't recognize your true ability. You are proud of what you have accomplished but feel you have not lived up to your full potential.* Some of the claims assert things many people *wish* were true of them. E.g., *You have quite a bit of unused potential just waiting to burst forth. You show excellent potential for leadership.* Some of the claims are rather specific but the reader is highly motivated to find significance and meaning in them and often does, even though the claims are just guesses or thrown out like bait to see if anyone will bite. E.g., *something about a red shoe and the name Michael; does that make any sense? He says he forgives you and it's time you moved on with your life. She says don't throw away that old calendar.*

Humans are very good at finding meaning or significance where there is none. Psychologists call the process of validating words, initials, statements, or signs as accurate and personally meaningful and significant *subjective validation.* Since most of us are innumerate, we often find significance in purely coincidental events. If you think of all the pairs of things that can happen in a person's lifetime and add to that our very versatile ability of finding meaningful connections between things in ambiguous situations, it seems likely that most of us will experience many meaningful coincidences, but we are the ones who give them meaning. Given the fact that there are billions of people and the possible number of meaningful coincidences is millions of billions, it is inevitable that many people will experience some very weird and uncanny coincidences every day. Put another way, with a large enough sample size, just about any possible weird coincidence will happen. This is sometimes called the *law of truly large numbers.* However, how likely a person is to attribute a dream that comes true to clairvoyance or precognition, for example, may be due to preconceived notions as well as to innumeracy. If one believes strongly in the paranormal, then one is likely to interpret an uncanny experience as paranormal. The biases and prejudices of one's worldview affect every perception and memory one has. If one's worldview is packed with nonsense and falsehoods, there is little hope that one can think critically about most things. If your disposition is to find psychic events everywhere you look, there is little hope that you will be willing or able to consider alternative explanations for what you take to be psychic events. Of course, if your worldview excludes the possibility of paranormal events, you will not be able to consider paranormal explanations for anything.

We're all socialized to conform to many ideas and practices. Many of us are also encouraged to ostracize those who are seen as different from us or who believe or behave in ways our in-group considers unacceptable. Thus, in some quarters, it may be rare and daring for a person to acknowledge belief in the paranormal. If you've been taught that psychic things are the work of the devil or the result of ignorance and superstition, you will probably not be inclined to favor paranormal explanations either for your own or for other people's experiences. Even so, our worldviews are often shaped by more than one social group and some of the "rules" of ostracism and shunning that we are taught are contradictory. We develop a hierarchy of rules based on which social group has the most influence on us. The anti-paranormal community is just one among many anyone might come in contact with while growing up. Consider, for example, that even though probably 99.9% of the biologists in America accept evolution as the essential foundation of modern biology, more than 50% of American adults reject evolution in favor of a view taught by Jewish and Christian fundamentalists: *God created all species at once a few thousand years ago.* Religious institutions have widespread influence in reinforcing the idea that creationism implies evolution is wrong. On the other hand, scientific and educational institutions do not have as much influence in reinforcing the idea that evolution is correct. This is partly due to the fact that many creationists are teachers and school board members and have influenced what is taught in our science classrooms. The ideas that the Earth is young and that all species emerged at once is believed by about half the population despite compulsory education and twelve to sixteen years of instruction, including science instruction. So, even though the vast majority of the scientific community considers belief in creationism and the paranormal to be unwarranted, the majority of citizens still prefer creationism to evolution and many still accept that paranormal events occur regularly.

Passionate Judgments

Few topics equal politics and religion when it comes to the power to arouse people's emotions to fever pitch, but other topics come close, e.g., paranormal claims, conspiracy theories, UFOs bringing aliens from other worlds, and "alternative" medicine. Many believers in the paranormal and their skeptical opponents are so emotionally committed to their beliefs that they are incapable of thinking critically about them. They tolerate no objections or questions regarding what they consider to be *overwhelming evidence* for or against such things as remote viewing for the military or telepathy in dogs and parrots. Strong passion is not necessarily a bad thing. If we didn't feel strongly about something, we probably wouldn't spend much time investigating it or thinking about it. But emotion can hinder the ability to think critically by preventing one from considering alternatives.

Judgments made while under the sway of a strong emotion, like anger or jealousy, are almost always regrettable. Decisions made while under heavy stress can be dangerous to oneself and to others. But when it comes to emotions hindering the ability to think critically, none can claim superiority to the emotion of *fear*. Nothing can motivate people like fear. Manipulators of beliefs and actions have known this for millennia. Fear can drive whole nations to engage in hysterical, irrational actions. Advertisers and phony healers use fear to sell us worthless products. Politicians play on our fears to manipulate us. For some, fear is the weapon of first resort. For others, it is the weapon of last resort. But fear is a weapon almost everybody uses. We often can't trust ourselves when looking for what is really motivating us to believe what we do or act the way we do. So we should ask ourselves: *Am I doing this because of fear? Do I believe this because I'm afraid not to? Are my fears justified?*

There is one fear, however, that has driven human beliefs and actions for as far back as we can reconstruct the history of human consciousness: *fear of the unknown*. Uncertainty drives this fear and has led us to thousands of superstitions and irrational actions. Undaunted, however, we have rationalized our fears and superstitions and irrational acts. We've created elaborate stories about them and these stories tell us a lot about ourselves, even if they don't really tell us much about the future or the nature of reality. The human brain is a story generator and the human being is a story lover. We have been making up stories for thousands of years and passing them on from generation to generation. There are stories of spirit serpents that pushed up the earth and created the rivers and mountains. There are stories of evil spirits manifesting as talking serpents to misguide and deceive us. We have stories for everything uncertain in our lives, everything we fear and everything we'd like to control. Where did we come from? No problem. We have thousands of stories to account for that. Death? There is no need to fear death, for we have thousands of stories of immortality. Why is there disease, sickness, bad fortune? We have stories for those, too. Why do bad things happen to good people? We have stories to explain that and anything else that puzzles you. Many of the ancient stories led to rituals being performed, partly to help remember the stories and partly to provide excuses for those times the stories didn't seem true. If the rain dance didn't work or reading the entrails of a sheep didn't produce the desired result, it was because the ritual hadn't been performed correctly or because we had offended some spirit. There is never anything wrong with the story. Any problems are due to us.

These stories and the practices that went with them eventually coalesced into the world's religions, which are reservoirs of humanity's superstitions and fears. That may not be all that religions are, but it is their essence. Religions are collections of stories and rituals that bind groups of people together. They are vats of magical thinking from which communities get their sustenance. Of course, occasionally one of the stories is actually correct and there is evidence that strongly supports the message of the story. For example, the Confucian account of the futility of identifying events in our lives as good or bad fortune is as true as any story gets. Who knows what is good or bad? What seems good today may seem bad tomorrow. What seems bad today may bring good tomorrow. But that story is one in a million. The stories of spirits in the streams and the mountains, sky gods, devils with horns and instruments of torture, words flying off paper to fill the air with divine thoughts, genocidal floods sent by the creator of the universe, gods incarnate as men or birds, eternal fires that never fully consume the flesh…these kinds of stories dominate in the telling and they dominate in the believing.

For many millennia there was support for many of our magical stories thanks to our psychological biases (like confirmation bias and the power of suggestion) and to unknown facts (like the facts that most diseases go away on their own or that touching a wound can infect it). Many stories, like those about the afterlife, were protected by the shield of being impossible to test. Then science arrived and it became possible to verify or falsify many of the old stories. But here we are more than a century after Darwin published his *Origin of Species* (1859) and a group of scientists organized the Society for Psychical Research (1882). The latter has achieved nothing of note, while the former provided the foundation of modern biology. Yet, more people accept as fact such things as spirit communication than accept the fact of evolution.

We love stories and our love for them is independent of whether they are true. This love of stories is natural. Encasing our stories in impenetrable sanctuaries is also natural; it is how we protect ourselves from doubt and fear. Many of these stories seem incredible to the rational mind, but even believers willingly acknowledge that fact. They revel in it. The more impenetrable and mysterious a story, the more value religious minds give to it. Belief in such things as the Trinity, incarnations of divinities, and stealing of babies by fairies and replacing them with decoys have the highest standing among their adherents. *I believe because it is absurd*, said Tertullian of his acceptance of Christianity. (Maybe he said *I believe even if it is absurd*.) I can't think of a better way to put it. But that is only half the story. The other half is that these stories, no matter how absurd, serve some useful purpose in the lives of billions of people. But those purposes are independent of the truth of the stories. To think critically, therefore, one must engage in the very unnatural act of challenging stories that the majority of people on the planet find useful and think are absolutely true and beyond challenge.

Of course, many religious people have no trouble thinking critically in areas that don't involve religion. They can do brilliant cost/benefit analyses of complex individual, corporate, or international problems. They can solve complex word problems in mathematics. They can do meaningful and significant scientific investigations. And, of course, many religious people can think logically as they draw inferences from their assumptions. Many of us, religious and non-religious, have cognitive blind spots, areas where we cannot think critically because our irrational biases make it impossible for us to be open-minded inquirers. But even in those areas where we are not driven by irrational biases, we remain vulnerable to multitudinous snares of deception and manipulation. One of the most powerful of those snares is language itself, a topic we now turn to.

SOURCES FOR CHAPTER THREE 270

IV
Extraordinary Renditions and Graphic Illusions in a Vaguely Familiar Universe

---

"A good catchword can obscure analysis for fifty years." — Wendell L. Wilkie

"...in our time, political speech and writing are largely the defense of the indefensible."—George Orwell

"If I turn out to be particularly clear, you've probably misunderstood what I've said."—Alan Greenspan

---

It should not come as a shock that some members of our species use language to manipulate thought and behavior. These are the doublespeak experts. They are the ones who send people off for extraordinary rendition *and* inhumane debriefings *without* kidnapping *or* torturing *them. Such people engage in* pretexting *but they never* fraudulently misrepresent themselves in order to get information about someone else. *Some people are so good at abusing language that they can redefine just about any evil so you won't recognize it. They are the* ethnic cleansers *who build* relocation centers *for the survivors of* collateral damage.

Manipulators use framing *to control the debate over many issues. One can mislead others with* pictures *and* diagrams *just as effectively as with words.*

Deconstructing the Language of Politics

When Idi Amin, the now-deceased former president of Uganda, created a squad of murderers to kill his political enemies, he named it The Public Safety Unit. When officials of the U.S. government denied that our country tortures suspected terrorists, they admitted that we do *intense debriefings that are sometimes inhumane.* Government spokesmen admitted that official enforcers from the C.I.A. or the military deprive captives of sleep, make them go naked for long periods, frighten them with vicious dogs, and dunk them in water to simulate drowning. The latter is called "waterboarding," a sonorous expression that might evoke visions of a day at the beach, which it assuredly is not. But our people do not *torture* anyone. How do we know this? We know this, we're told, because *we* are doing it and *we* don't torture. Former C.I.A. director Porter Goss put it this way: "*we don't torture, we do debriefings.* Torture doesn't get results. We get results with our methods." In any case, while the economy is in a period of underemployment, you might rethink applying for that "risk management" job. It may not be in the financial sector, and if hired you may find yourself working as a mercenary in Libya and be expected not to flinch during *refined interrogation techniques at a black site.*

Not all examples of doublespeak are as wicked as calling murderers *public safety officers* or calling torture *debriefing* (or genocide *ethnic cleansing,* a concentration camp a *relocation center,* indiscriminate killing of non-combatants *collateral damage,* ad nauseam.) When Newt Gingrich, former Speaker of the House of Representatives, wanted to arouse sympathy for the Republican Party prior to the 2006 November elections he created a new political enemy: "San Francisco left-wing activists." If you haven't heard of these folks, they are the ones who have "San Francisco left-wing values." These are people with *liberal values* who run with the *elite media.* They want to raise taxes and either "cut and run" or "run and hide" from Iraq. They represent "the failed policies of higher taxes, more regulation and bloated bureaucratic structures of the past." They support policies of *appeasement* and *defeatism.* According to Mr. Gingrich, "If you think you have too much money in your family budget, then you have a party to vote for, because Democrats will gladly raise your taxes shifting your money from your family to Washington bureaucrats." And, "if you want to go back to high taxes, high interest rates, high inflation, slower economic growth, more unemployment, fewer savings, shorter vacations and more bureaucracy, then you have a party in the Democrats." Shorter vacations? Yes, according to Mr. Gingrich, the Democrats can even shorten your vacation if elected. That's how evil and powerful those San Francisco liberal, radical, left-wing Democrats are. Gingrich must have a short memory, however, because the

Democratic Clinton years were the best economic years in decades and the violent crime rate went down dramatically during those years of a booming economy. I'm not saying Clinton or Democrats were responsible for those good things, but I do know that it was not an era of high taxes, high interest rates, high inflation, slow economic growth, more unemployment, and shorter vacations. If there were fewer savings during the Clinton years, it may have been because people remembered what happened to their savings during the Reagan years when Republicans were raising their glasses to "the Gipper" as the savings and loan industry was deregulated. Many elderly people lost their life's savings thanks to that bit of government deregulation. In any case, I know for a fact that my vacation time didn't get any shorter when the Democrats were in power and I don't think anybody else's vacation time was affected, either.

There's very little that Idi Amin and Newt Gingrich have in common, but they both used words to shape how people think about things without regard for the truth. Amin wanted to gloss over his evil deeds, while Gingrich wanted to arouse voters to support Republicans by painting their main political opponents as enemies of the people. He was especially concerned with getting the vote out because of some recent bad news for the nation, which, when seen through partisan eyes, meant bad news for Republicans in upcoming elections. After months of leaks, the Bush administration released parts of a classified National Intelligence Estimate that stated that in the opinion of 16 distinct spy services inside the government the overall terrorist threat has *grown* since the September 11, 2001 attacks. In short, the report said that U.S. policies to restrain Islamic radicalism had failed. Rather than being in retreat, Islamic terrorism was on the rise. More bad news came with the publication of a book by Bob Woodward called *State of Denial: Bush at War, Part III*, a derisive commentary on the Bush administration's handling of the war in Iraq. Then, as if to prove the old adage that *bad things happen in threes to people who can't count higher than three*, a scandal emerged involving Republican congressman Mark Foley. He admitted that he'd sent sexually explicit e-mails to House pages who by law are considered children because they are under the age of eighteen. Foley's behavior was especially outrageous because he had sponsored a federal Internet predator law that makes it a federal crime to solicit, discuss, or allude to anything sexual on the Internet with a person under the age of 18. The congressman resigned and went into rehab, putting some of the blame for his bad behavior on alcohol, homosexual desires, and being abused as a child by a priest. Gingrich and other Republicans would have liked the press to have covered this story and any other negative stories involving Republicans quickly and discreetly before moving on to less embarrassing topics. The press did not oblige, leading Gingrich to comment: "The elite media are giddy with anti-Republican euphoria. Their coverage has not been this biased against Republicans in three decades." Can anyone say *hyperbole*?

Gingrich didn't really expect the news media to ignore the war in Iraq, the National Intelligence Estimate, the Woodward book, or the Foley story, or to apply a bizarre interpretation of the "fairness doctrine" and dig up equally embarrassing stories about Democrats. He chose his words to arouse indignation and get voters to choose Republicans in the upcoming election. By framing the issues the way he did, Gingrich tried to turn criticisms of the Bush administration and a scandal involving a Republican into *a biased, elite media* issue. Gingrich also tried to arouse as much fear as possible. His reference to "left-wing San Francisco radical ideas" was aimed at Nancy Pelosi, a Democrat from California's 8th District (which includes most of San Francisco) who was elected to the position of Speaker of the House of Representatives when the Democrats captured enough congressional seats in the election to give them a majority, despite the efforts of Gingrich and other Republicans. Gingrich didn't say anything about it but polls were also showing that a majority of the American public did not agree with President Bush's Iraq policy. But why address the issues when rhetoric is cheaper and a proven effectual weapon of mass destruction? Gingrich probably doesn't even appreciate the irony of trying to manipulate the media and the public with his words while berating the media for their unfairness and bias. Only Vice President Cheney approaches Gingrich for chutzpah with regard to media fairness. Cheney complained loudly of bias in the media during an interview with Armstrong Williams, who was being paid by the Bush administration to write biased news stories favorable to its policies. Mr. Gingrich and Mr. Cheney probably have no more concern for fairness in the media than Idi Amin had for public safety.

Language can be a powerful weapon for manipulating thoughts and actions. Newt Gingrich is a master with words, but even people who are barely competent speakers of their native tongue can learn how to rant and rave and arouse sympathy. Witness John Doolittle, Republican ex-Congressman representing California's 4th District, in a debate with his Democratic challenger, Charles Brown. Doolittle repeatedly referred to Brown's membership in the American Civil Liberties Union and called him a "flimflam man" with "extreme liberal views" who appeared at an anti-Iraq war rally with Sean Penn and Cindy Sheehan (an anti-war activist whose son, U.S. Army Specialist Casey Sheehan, was killed in Iraq). "I'm a conservative," said Doolittle. "Tell us who you are. Don't be a flimflam man who is conservative in the daytime and hobnobs with Willie Brown (former California Assembly leader and mayor of San Francisco) and the liberals in San Francisco at night." Doolittle also suggested that Charles Brown was supporting a notorious sex ring by belonging to the ACLU. Brown didn't remind Doolittle that the ACLU defended conservative radio star Rush Limbaugh when he was under investigation for violating drug laws. In 2003, Limbaugh was investigated for illegally obtaining the prescription painkillers oxycodone and hydrocodone. He admitted he was addicted to the drugs, but was not charged with a crime. The ACLU filed an amicus curiae brief in Florida on behalf of Limbaugh and argued that state officials were violating his right to privacy in seizing his medical records for their drug probe. In any case, Doolittle knows that Brown doesn't have to agree with everything the ACLU does, even if he's a card-carrying member of the organization. His association with that organization is irrelevant but it packs a mighty emotive wallop with people who don't think critically about the matter.

Brown's supporters showed up outside the debate location with placards supporting their candidate. Doolittle referred to them as "screaming hooligans." Brown, for his part, accused Doolittle of taking money from Mark Foley and "convicted felons" (lobbyist

Jack Abramoff) and of shirking his duty during the Vietnam War by doing missionary work for the Mormon Church instead of fighting for his country as Brown and other Mormons did. The lowlight of the night came when Doolittle accused Brown of being "totally out of touch" and then proceeded to claim that the Iraq war, which Doolittle supports, "leads to what the Bible ultimately says" about Armageddon. Apparently Mr. Doolittle thinks U.S. foreign policy should be aimed at fulfilling Biblical prophecies and that belief puts him in touch with something.

Neither candidate bothered much with the issues. They were too busy trying to use words to portray their opponent as either an immoral, felonious chicken-hawk or as a liberal supporter of pedophiles.

Why do politicians do this? The short answer is that it is effective, cheap, and easy. The long answer was given by George Orwell in his 1946 essay "Politics and the English Language," where he wrote—among many other incisive things—that political speech is "largely the defense of the indefensible." As a result, "political language has to consist largely of euphemism, question begging and sheer cloudy vagueness." Such language requires people to think less than if it were detailed, specific, and clear. Expressions like "liberal" are so emotively charged that no explanation is expected or required. Such expressions are used like weapons to attack an opponent without having to be specific about anything. They require no thinking on the part of those whose support you are trying to rally. Politicians use slogans like "cut and run" to describe any proposal to get out of Iraq, "stay the course" for ignoring any such proposals, and "flip flop" for changing one's mind in light of new evidence. They speak this way to relieve themselves and the hearer or reader of the responsibility of thinking about anything. Such expressions have little cognitive content but they carry powerful emotive currents. "Cut and run" and "flip flop" sound weak and harmonize with words like "appeasement" and "defeatism," while "stay the course" sounds solid and strong and is associated with words like "victory."

When I started college in 1963 "liberal" was used primarily with a positive connotation and "conservative" had a slightly negative connotation. Barry Goldwater, often credited with founding the modern conservative movement in American politics, was trounced by Lyndon Johnson in the 1964 election. Goldwater was called a conservative by many people who meant it in a negative way. I don't recall anyone calling Johnson a liberal. Now, whenever Republicans want to evoke a negative feeling about a person or an idea all they have to do is use the word "liberal." Over the past forty years, Republicans have managed to get the public to associate the word "liberal" with increases in taxes, government spending, government regulation, and uncontrollable growth of government bureaucracy. Republicans have also managed to persuade a large segment of the general public that all government spending and regulation is bad, that any tax increase is bad, and that government is the root cause of most of our evils. The evidence to support these views has never been provided by the Republicans, but that fact does not seem to bother many of our fellow citizens who are convinced of its absolute truth. This is quite an amazing feat when one considers that President Ronald Reagan, the hero of the anti-liberal establishment, increased taxes and government spending by 68% in his first six years in office (1980-1986), while doing very little to stop the growth of government regulation and bureaucracy. The U.S. government spent $591 billion in 1980. It spent $990 billion in 1986. Federal spending as a percentage of the gross national product in 1980 was 21.6%; six years later it was 24.3%. The deficit rose from $60 billion in 1980 to $220 billion in 1986, while the national debt more than doubled from $749 billion to $1.7 trillion. Under "conservative" George W. Bush, the national debt rose to about $9 trillion. There was no deficit when President Clinton left office and George W. Bush took over. Within a few years, the deficit was over $700 billion. According to USdebtclock.com, under President Barack Obama the deficit has risen to more than $14 trillion on 28 October 2011. Yet, we still hear many Republicans talk about *the return* to fiscal conservatism. That would be a return to the Clinton years, which is not exactly what the Republicans want.

Furthermore, the loudly proclaimed tax cut of 1981 was little more than an illusion. Tax rates for higher-income brackets were cut but the average person saw his or her taxes rise for various reasons. One reason had to do with moving into higher tax brackets. Another had to do with increases in social security taxes. But the main reason taxes for the average person kept increasing under President Reagan was that the increases were given deceptive names like "revenue enhancements," user and licensing "fees," and "plugging loopholes." It wouldn't be long before these terms were joined by "rate adjustment," "benefit reduction," "service charge," and other euphemisms for *tax increase*. As a result, federal tax receipts increased from $517 billion in 1980 to $769 billion in 1986, an increase of 49%. I won't bother detailing how the U.S. economy went into its greatest "recession" since the Great Depression after years of deregulation of banks and other financial institutions. Yet, despite the obvious fact that reducing government regulations on businesses and lowering taxes on corporations was followed by a downward spiraling economy, Republicans still run for president of the United States by calling for less government and lower taxes. Why? The *words* resonate with a large segment of the voting public.

What happens in the real world is independent of what goes on in the world of public discourse. "Liberals" were blamed for the increases in government spending and taxes. Democrats were identified as liberals. Republicans framed themselves as supporters of tax relief and tax cuts. If Republicans could increase government spending and the federal deficit while raising taxes but convince a substantial part of the general public that they were doing otherwise, what couldn't they accomplish? They went even further and convinced their followers that they were there to protect the public against the evil liberals who *would* raise their taxes, bloat the bureaucracy and do all kinds of other bad things (like weaken our defense by cutting military spending). Apparently, it doesn't matter what is true or false. What matters is how you frame the issues.

In 1963, conservatives were identified by their disdain for the growth of the federal government in its apparent attempt to become all things to all people. Programs like Social Security and anything resembling a Franklin D. Roosevelt New Deal program were suspect. Nor did conservatives want the federal government interfering in such things as civil rights. Politicians like Richard Nixon and Barry Goldwater were vehemently opposed to the

Supreme Court decision in Brown v. Board of Education because they thought that segregation and other racial issues should be left to the states to work out. Conservatives didn't want Congress or the Supreme Court telling the states how to treat their citizens. But Goldwater also didn't think the federal government should be telling women that they can't have an abortion. His attitude on gays in the military was that it was no big deal as long as the soldier was ready to make war not love. By the time Senator Goldwater retired for good in 1987, many of his positions were considered "liberal." Conservatives now defend the federal government telling the states what to do. They've supported constitutional amendments to bring prayer back to public schools, to ban burning the flag in protest, and to forbid same-sex marriage. And they've tried to stack the Supreme Court with justices who they think might vote to overturn Roe v. Wade. What is amazing is that the conservatives can continue to support such things as government interference in what areas scientists can do research (e.g., stem cell and cloning research), while representing themselves as opponents of big government and defenders of getting government off our backs. They are able to succeed at this bit of legerdemain in part by loudly opposing government regulation of industry and commerce in the name of keeping the economy growing so unemployment and taxes will stay low. If that happens then more people will have more money and fewer people will need help from the government, which is good, good, good.

I have focused on Republicans and their successful campaigns at framing political issues simply because the Republicans have been so successful and the Democrats have failed miserably at controlling the language of political discourse. What has happened to political discourse makes it possible for people to feel comfortable in their parties without having to do much thinking. Political campaigns are not mainly about issues but about words. They're not mainly about ideas but about the emotions and images that words can arouse. One good catchword like "liberal" can obscure analysis for decades. Unfortunately for the Democrats, they don't have any go-to catchwords to hurl at their opponents. "Right wing ideologue," "radical right," or "religious right" just don't have the framing power that "liberal" does.

After Barack Obama was elected president and followed in the footsteps of George W. Bush by continuing the bailouts of corporations and financial firms in an effort to save the economy from total disaster, he has been labeled a "socialist" or a "communist." Nobody called Mr. Bush a socialist, even when he was at his Orwellian best, claiming he had to abandon free market principles to save the free market. Whether one agrees with these bailouts, they are light years away from socialism or communism. Such words might have worked at one time on a large segment of the American public, but today they are probably not going to be as effectual as the campaign to bloat the word 'liberal' with expanding negative emotive and cognitive content. In any case, the bailouts were started by a Republican who thought they were necessary to save our "free" economy. The idea of loaning or giving huge sums of money to private enterprises so that we can continue to enjoy a free economy seems contradictory on its face, but delving deeply into this issue is beyond my pay grade. Every rational person who knows only a smidgen of economics knows that capitalism in the U.S. is not a totally free economy. There have always been some restrictions on trade and commerce. The only issue reasonable people can disagree on in this area is *what* those restrictions should be. Anyone who claims there should be *absolutely no restrictions* on trade and commerce is not talking sense.

Words to Go to War With

Politics is not the only arena where discourse has been co-opted by wordsmiths rather than thinkers willing to engage in rational discourse and conversation. In 1993, the Pentagon denied it had rigged a test of its Strategic Defense Initiative (SDI) or "Star Wars" project. Officials admitted that the target was artificially heated so that it appeared ten times larger to a heat-seeking missile, but they insisted that the target missile was "enhanced" not "rigged." Whatever else the Pentagon did with the results of this test, you can bet they used them to *enhance* their evaluation of the success of the program and *enhance* their budget requests. These are the same people who, with a straight face, can refer to a missile that blows up on the launching pad as *an incomplete success.*

The Department of Defense is notorious for its invention and use of euphemisms to disguise ugly realities. Body bags become *transfer tubes.* Bombing missions become *visiting sites* and targets are *visited* or *degraded, neutralized, cleansed, sanitized,* or simply *taken out.* But we never bomb anyone. Nor has the U.S. ever tried to *overthrow* a government. We've financed, advised, and otherwise promoted *nation building* and a few *regime changes,* however.

During the Reagan administration, Lt. Col. Oliver North was running a government within the government. He was selling arms to Iran at inflated prices and using the profits to support an insurgency in Nicaragua. North referred to the profits as *residuals* and *diversions.* When he destroyed the records of his activities, he described his actions as *cleaning up the historical record.* Even so, North doesn't come close to the egregious doublespeak of those who describe genocide as *ethnic cleansing.*

The kind of language abuse we are discussing here involves using euphemisms to manipulate how people think about things. If we can use dull language we are not likely to arouse emotions. If we don't arouse emotions we are not likely to arouse thinking about what we are talking about, which is exactly what the manipulators want. If people don't think about what the manipulators are saying or doing, they won't be objecting to their behavior. If I say we dropped a bomb on civilians and killed hundreds of children and old men and women, people might get upset and object that that is not a morally right thing to do. If, however, I say that a site was *visited* and there was some *collateral damage,* even if people know what I mean, they are not going to react very strongly to my words. The words are too dull to arouse much feeling.

Who will object if I admit *to extraordinary rendition for purposes of intensive debriefing*? You can't object unless you know what I mean and what I mean by such an expression isn't clear. This is jargon for kidnapping people and taking them to foreign countries where our agents or allies will torture them. Some citizens might object to their government engaging in such behavior, so it behooves the government to find words that aren't clear to describe their actions.

If the government calls a law The Clean Air Act it is almost a sure thing that the air will get dirtier. Anything called The Tort Reform Act or Class Action Fairness Act is unlikely to have any interest in tort reform and is probably designed to protect corporations from being sued while protecting the right of corporations to sue as they see fit. Feminists for Life is a group that seeks to outlaw all abortions, no exceptions, not even to save the life of the pregnant woman. The Fairness in Asbestos Injury Resolution Act (the FAIR Act!) was a bill sponsored by Patrick Leahy, the highest ranking Democrat in the Senate, and Arlen Specter, Republican and Senate Judiciary Committee chairman. They claimed the bill would help the victims of years of abuse by mining companies, but the goal of the law was to make sure that those companies didn't get sued too badly for all their wrongdoing. The bill's main purpose was to limit the amount of compensation victims could receive.

In 2002, Frank Luntz wrote that "The most important principle in any discussion of global warming is your commitment to sound science." If you didn't know better, you might think he was a scientist advocating rigor in discussions regarding the effects on the global climate from the increasing abundance of carbon dioxide in the atmosphere. In fact, he's a Republican strategist and he was advising Republican congressional candidates on what words to use when talking about climate change. The phrase "global warming" is misleading because the warming effect isn't evenly distributed. Colder, dryer places will be more affected than warmer, wetter places. And the average global temperature rise is only a fraction of a degree. "Sound science" is a catchphrase used to undermine the consensus of the scientific community about any issue, including climate change. The expression seems to have been created by APCO Associates, an industry front group designed to label environmentalists as purveyors of "junk science." In short, what most other scientists call sound science, APCO calls "junk science." How could these linguistic alchemists turn sound science into junk science? It was actually quite easy, a lot easier than turning lead into gold. You claim that "sound science" requires an endless amount of further research because we're never absolutely certain about anything. There's always one study here or there that indicates DDT or smoking improves memory or that there is no climate change going on that is of any interest. We had better do more studies to make sure. But no matter how many studies you do, there is always the possibility that the next study will prove you wrong, so you had better do more studies. And no matter how good the studies are, no study is perfect and critics will always find some flaw that they can focus on, magnify, and use to prove their point that even the best science is junk science in disguise. In short, by delaying action until the endless studies are done or the perfect study arrives, no restrictive action is taken, which is exactly what many businesses and industries desire.

Thinking can also be manipulated by the positive or negative attitude words express. *Freedom fighters* are the insurgents you support. That's what President Reagan and Oliver North called the Contras, who they supported in an attempt to overthrow the government in Nicaragua. Those in power in Nicaragua might have called them *rebels, insurgents, terrorists,* or *guerillas.* On the other hand, *the resistance* refers to the rebels you support. Insurgents anywhere probably think of themselves as the resistance, liberators, and freedom fighters. The words we use in such situations tell us more about our feelings and beliefs than they do about the reality of what we are using the words to describe.

Some words look and sound like they are conveying information about the world, when all they do is express a positive or negative attitude and stir up similar feelings in those who read or hear them. On one side are what Jamie Whyte calls *hurrah words*: *peace, love, victory, happiness, security, safety, protect, innocent, freedom, liberty, justice, democracy, courage, confidence,* and *tax relief.* If you're trying to arouse sympathy to your viewpoint, no matter how obnoxious, deceptive, or pernicious that viewpoint might be, sprinkle your speech with plenty of hurrah words. On the other side are the *boo words.* If you're trying to arouse opposition to others be sure to include several boo words in your speech: *hate freedom, hate liberty, terrorize, attack, barbarity, murdered, threat, cowards, evil, kill, extremists, radical,*

tyranny, dictator, arrogant, woo-woo, pseudoscience, and *liberal.*
Boo words arouse sympathy by provoking contempt.

Visual Misdirection

Images can be even more effective than words at arousing sympathy or disgust for a position. One good image for your side can overthrow ten good arguments from your opponents. Show a photo of a bruised and bandaged woman next to her smashed up car and launch into scary tale of "road rage" or show pictures of a dozen sad looking women or children suffering from cancer or neurological disorders and launch into a scary tale of the dangers of silicone breast implants or vaccines. It will not matter that the scientific evidence indicates that so-called road rage is not a major problem or that the scientific evidence overwhelmingly demonstrates that breast implants do not cause any major diseases and that vaccines do not cause autism. Boo pictures work just like boo words: they make it unnecessary to provide proof for your position. You can bypass thinking altogether with hurrah pictures, too. You don't have to fear tigers; look at this photo of little Beatrice cuddling up to a 300-pound Bengal. Smoking can't be that bad for your health: look at these two businessmen lighting up cigarettes after a little game of racquetball and then look at this mountain climber who has just reached the top of the Alps and lights up a menthol cigarette! What pleasure! And look at that manly cowboy riding off into the sunset with a Marlboro in his mouth! (Who can forget the boo picture of the Marlboro man riding with his oxygen canister strapped to his horse's saddle?)

Graphics can be misleading also. We're all familiar with the deceptive bar chart that positions a short stack against a much taller stack to create the illusion of great change. The two graphs shown here (**figure 3** and **figure 4**) might be used to represent the same data. Figure 3 shows larger units on the vertical axis, creating the impression of a larger difference between the two columns than the chart in figure 4, which uses smaller units on the vertical axis. Both graphs represent data in which the right column represents a 10% increase over the left.

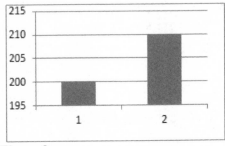

Figure 3

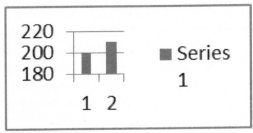

Figure 4

Such deception might cause some minor confusion, but some graphics can be deadly according to Edward Tufte, the author of *Visual Explanations*. He suggests that if Thiokol engineers had had better charts, their arguments made to NASA the night before the fatal *Challenger* mission would have prevailed and the launch would have been delayed. Seven crew members died when the space shuttle exploded because two rubber O-rings failed due to the chilly temperature. The engineers provided thirteen charts to NASA as part of their evidence that the threat of cold to the O-rings strongly indicated that carrying out the mission—when temperatures were predicted to be in the 26°F to 29°F range—would lead to failure of the O-rings. Tufte reproduces some of the Thiokol charts in his book and provides charts of his own that he thinks clearly show that cold temperatures significantly compromises O-ring stability. Tufte is right about the engineers' charts: they do not show a clear relationship between temperature and O-ring damage. However, his charts aren't much better on that count. His charts are clearer and aren't mucked up with superfluous pictures of rocket ships, as are some of the

Thiokol charts. Tufte's charts rely on the same data that the Thiokol engineers charted: temperature at the time of previous space flights. However, there had never been a space flight when the temperature was in the 26°F to 29°F range. The lowest temperature at the time of a previous launch was 53°F. Even though Tufte's charts show that there were more O-ring damage incidents at 53°F than at any other temperature, they show damage on two occasions when the temperature was 70°F. For 24 launches, there were only eight "erosion incidents" and even though the four lowest temperatures (ranging from 53°F to 63°F) had the most problems, the graphic is not that compelling to warrant making assumptions about O-ring failure at lower temperatures. Only with the benefit of hindsight does Tufte's argument seem to have merit. What was needed was a graphic that illustrated numerous tests for erosion done on O-rings at different temperatures independently of space shuttle launches. A graphic that would have been compelling would have looked something like the following:

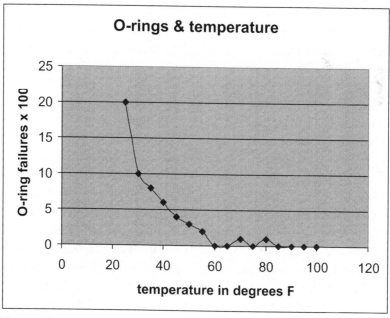

Figure 5

This graph clearly shows that at lower temperatures, O-rings start to fail at a significantly higher rate than at higher temperatures. Tufte's graph looks like the above graph (**figure 5**), except that the left half was extrapolation and not based on actual data since no launch had ever taken place when it was below 53°F.

Psychic Ambiguity or Clairvoyance in the Dark

While it is possible to modify the visual presentation of data to clarify its implications, some language is so vague, ambiguous, or obscure that its meaning can't be determined until after something has happened. For example, fans of the 16th-century French astrologer Michel Nostradamus claim that he predicted the *Challenger* disaster long before the Thiokol engineers had prepared their charts. Here is what Nostradamus wrote 500 years ago:

> From the human flock nine will be sent away,
> Separated from judgment and counsel:
> Their fate will be sealed on departure
> Kappa, Theta, Lambda the banished dead err.

Or, in modernized French:

> D'humain troupeau neuf seront mis à part,
> De jugement & conseil separés:
> Leur sort sera divisé en départ,
> Kappa, Thita, Lambda mors bannis égarés.

How, you might ask, did anyone find mention of the *Challenger* disaster in these four lines of apparent gibberish? They find them the same way some people find the Virgin Mary in a burnt tortilla or violent sword fights in an inkblot: they use their creative imagination and shoehorn their beliefs to the situation. Thiokol made the defective O-ring that is blamed for the disaster. The name has a 'k,' 'th,' and an 'l.' Never mind that there were seven people who died, not nine. The rest is vague enough to retrofit many different scenarios. Who, you might ask, would pay any attention to the obscure ramblings of somebody who's been dead for half a millennium? Many people, apparently. In 2001 on my Skeptic's Dictionary website, I was averaging a little under half a million hits a week on hundreds of articles I had posted. The week after the 9/11 terrorist attacks, the Nostradamus entry alone received about 4 million hits.

As noted in the previous chapter, the ability to find meaning and significance where there is none is something humans are very good at. Retrofitting vague or obscure words or symbols to events is common among so-called psychics or prophets. Such people might truly be useful if they could predict the future before the future occurs, but unfortunately for us, their powers seem to be mucked up so that they can clearly predict the future only when the future has become the past. Self-proclaimed psychic detectives, for example, never pinpoint the criminal at the start of an investigation. Only after the investigation is over does their foresight become apparent. For example, Court TV brought in psychologist Dr. Katherine Ramsland to explain how psychic detective Greta Alexander helped solve a case:

> Police needed a body, but he (the confessed killer) wouldn't reveal the information, so they turned to a psychic, Greta Alexander. She said that a body had been dumped where there was a dog barking. The letter 's' would play an important role and there was hair separated from the body. She felt certain the body was in a specific area, although searchers found only a dead animal. She asked to see a palm print of the suspect—her specialty—and the detective brought one. She said that a man with a bad hand would find the body. Then searchers found a headless corpse, with the head and a wig nearby. The man who found it had a deformed left hand. There was water nearby.

Dr. Ramsland would like us to believe that Alexander got psychic impressions from some other dimension (supernatural or paranormal) by some mysterious process. Another way of looking at this detective's work is to see it as revealing nothing of use that led to the discovery of the body. Only after the body had been discovered was it possible to put an interpretation on Alexander's statements that seems to fit the situation. But did it? Is Dr. Ramsland seeing patterns and finding meaning where none exists? She doesn't seem bothered by the stretch it takes to get from "hair separated from the body" (which covers a multitude of possible scenes) to "headless corpse, with the head and wig nearby." Head and wig near the body are seen as hair separated from the body. But a million things could be seen as examples of "hair separated from the body." Those of us who don't have intermittent access to the paranormal might think that an image of a beheading would surpass in power an image of hair separated from the body. Also, there is no limit to the number of words that can be related to the letter 's'. At any given moment in most cities, many dogs are barking. Some of them might even bark if they stumble upon a headless corpse. There's water near all habitable places. But one reason many believers won't accept that they are retroactively creating the meaning of Alexander's utterances is that they are impressed by the specific nature of some of her claims. "Bad hand" is not as ambiguous as the letter 's' or the claim that a body will be found in a shallow grave or near water. Such specificity, they think, eliminates guessing or luck as a reasonable explanation for alleged accuracy. However, "bad hand" is ambiguous and it too could refer to many different things, including "deformed left hand of the person finding the body." A person with arthritis has a bad hand. A physician whose writing looks like chicken scratch has a bad hand. A gambler who's lost all his money had a bad hand. In any case, the vision of the deformed hand of the one finding the body is worthless information, except for those playing the game of "let's see if we can make sense out of this gibberish."

Alexander did a lot of self-promotion, but some psychics get assistance from scientists of dubious reputation. Dr. Gary E. Schwartz of the University of Arizona declared that he became convinced that Allison DuBois was psychic when she claimed to get a message from a dear friend of Schwartz who had died recently. The deceased told DuBois, she said, to share the message: "I don't walk alone." To most of us that might seem like an innocuous bit of information akin to "here in Heaven I'm playing bridge with all my friends and none of our pets have fleas." Schwartz saw it differently. "My friend had been confined to a wheelchair in her last years," he said. "There is no way Allison could have known that." Even if she did know it, she didn't say that. Schwartz provided the wheelchair meaning, but he could have provided dozens of other interpretations that might have made sense to him as well.

When parapsychologist Dean Radin wanted to provide a convincing example of the success of the dream telepathy experiments at Maimonides Medical Center done by Montague Ullman and Stanley Krippner, he selected a case that demonstrates the problem of doing experiments that allow for the interpretation of ambiguous data to fit or not fit your hypothesis. He mentions two cases that involved a sender concentrating on a target (Max Beckmann's painting "Descent from the Cross," which depicts Jesus being taken down from the cross) and a sleeping receiver whose dreams are supposed to be influenced by the sender. The receiver is hooked up to a machine that reveals when he is in REM sleep and probably dreaming. At that point, the sender tries to affect the sleeper's dreams. In this case, the sender was given some visual aids to work with: a crucifix, a Jesus doll, nails, and a red marker. He was given instructions to nail the doll to the crucifix and use the marker to color the body with blood. One sleeper reported that he had dreamed of a speech by Winston Churchill and another reported that he had dreamed of a "native ceremonial sacrifice." There was a reference to "sacrificing two victims," something about "destroying the civilized," and "the awe of god idea." There is nothing in either dream of the crucifixion in all its gory representation, yet Radin considers these dreams successful telepathic "hits." Radin comments on the symbolic significance of "church-hill." According to legend, Jesus died on a hill and a church is named after Jesus who is considered by many to be the Christ. Furthermore, the crucifixion is looked at as a sacrifice by Christians, Jesus is believed to be both man and God (two victims?) in some traditions, and there was a god of some sort mentioned in one of the dreams. On the other hand, there are literally thousands of items to which one might retrofit these dreams. Allowing loose and symbolic connections to be made after the fact in order to evaluate the accuracy of the telepathy is more a measure of the cleverness of the judges than of the paranormal powers of the participants.

By Their Words You Will Know Them

Psychics and their defenders are not the only ones who bank on our help in finding their words meaningful and prophetic. Advertisers also require our assistance to give meaning to weasel words and other vague or meaningless expressions. Advertisers tell us that their products *help* or *may help* us; we understand them to mean that their products *will improve our condition*. They tell us that their product is *safer*, *lasts longer*, or has *20% more*; we understand them to mean that their products are safe, last a long time, and provide good value. I once bought a package of almonds that had "15% more" written across the packaging. I counted the almonds and calculated that the product contained *three* more almonds than it used to. I understood why the package didn't say "contains 3 more almonds!"

Sale! is another word advertisers use but for which we provide the meaning: we're going to save money! When a sale sign says you can save "up to" 50%, the hope of the one having the sale is that you will think you're going to get things for half price. The weasel phrase "up to" means you won't save *any more* than 50%. The store with the sale might even give you a *warranty* on the sale item. Many years ago I bought a new car with a three-year warranty on the battery, which died after two years. I carried the dead battery to the counter at the car dealership and was told that according to the conditions of the warranty I had to drive the car in. Driving a car with a dead battery would require powers that I don't possess. The dealer did note that I could have my car towed in but that the towing would probably cost more than a new battery. Furthermore, he told me, the warranty was pro-rated so that the most I could hope to collect for my old dead battery would be about one-third its value. It turned out that even with a 33% discount on a battery from the dealer, it was still costlier than a new battery from an auto parts store. To use my warranty would have cost me quite a bit of money. I no longer pay much attention to my automobile warranties. Nor do I pay heed to sale signs that say *only* $9.99. Sounds like a deal, though. A friend of mine paid $10 for the same thing. Boy did he get ripped off!

Advertisers know which words are likely to push our buttons. *Hurry! Limited time only! Save!* They know that many of us will give these vague words a precise meaning favorable to them. Some journalists also use words that are likely to push our buttons. For example, some journalists will refer to "racial preference programs," while others will refer to "affirmative action programs." Some refer to the "anti-abortion" movement, while others refer to the "pro-life" movement. It is unlikely that the writers are unaware of the different responses these terms evoke in most people. A 1992 Louis Harris poll found that 70 percent said they favored "affirmative action" while only 46 percent favored "racial preference programs." Journalists can appear to be neutral when using such expressions, but they are sneaking in their opinions. If I want to evoke sympathy, I use the term most people respond positively to. If I want to evoke disapproval, I use the term that fewer people respond positively to. I appear unbiased because I don't use expressions like "the liberal push for racial preference" or the "promotion of fairness and justice by affirmative action." We will read about "gay marriage" and "sexual preference" in one publication but "same-sex marriage" and "sexual orientation" in another. The language seems innocuous enough, but it's loaded with subtle connotations. Most of us recognize the significant difference in emotive content when describing people as suicide bombers or terrorists rather than as martyrs. But referring to almost every bombing attack on Palestinians as *retaliation* is a much more subtle way of judging the news for us by indicating that the Israeli action was justified because they were provoked. Of course, the word 'retaliation' can be used only because journalists select for reporting an arbitrary slice of time from a decades-long struggle. Now we are moving away from bias in language to bias in the media, a subject that deserves a chapter of its own. But first we will discuss another kind of bias that drives people to both irrational thinking and irrational behavior: groupthink.

SOURCES FOR CHAPTER FOUR   271

V
Driving an Edsel to the Bay of Pigs

"I really do believe that we will be greeted as liberators. I've talked with a lot of Iraqis in the last several months myself, had them to the White House....The read we get on the people of Iraq is there is no question but what they want to get rid of Saddam Hussein and they will welcome as liberators the United States when we come to do that." —Dick Cheney, March 16, 2003

"In Iraq, a ruthless dictator cultivated weapons of mass destruction and the means to deliver them. He gave support to terrorists, had an established relationship with al-Qaeda, and his regime is no more." —Dick Cheney, November 7, 2003

"Dick Cheney says he'd invade the wrong country—again." —David Letterman, August 30, 2011

Why is it that groups of highly intelligent people often make bad decisions? What do companies have in common when it comes to making good group decisions or bad decisions? In this chapter, we'll examine groupthink.
We'll also explore the negative effect of communal reinforcement *on critical thinking. Communal reinforcement is the process by which a claim becomes a strong belief through repeated assertion by members of a group or community. The process is independent of whether or not the claim has been properly researched or is supported by empirical data significant enough to warrant belief by reasonable people.*

Groupthink and Deadly Decisions

On March 20, 2003, the United States led an invasion of the sovereign country of Iraq. President George W. Bush convinced the U. S. Congress and a good part of the American people that this act of aggression was necessary to defend the United States against terrorist attacks with weapons of mass destruction, including nuclear weapons. The argument for attacking Iraq had been presented by Secretary of State Gen. Colin Powell to the United Nations Security Council a few weeks before the invasion on February 5, 2003. Saddam Hussein possessed or was actively pursuing the materials needed for biological, chemical, and nuclear weapons of mass destruction. He also had ties to Osama bin Laden, Abu Musab Al-Zarqawi, and al-Qaeda, the individuals and the organization the U.S. had identified as planners and supporters of the terrorist attacks on several sites in the United States on September 11, 2001. Powell told the Security Council:

> We know that Saddam Hussein is determined to keep his weapons of mass destruction; he's determined to make more. Given Saddam Hussein's history of aggression, given what we know of his grandiose plans, given what we know of his terrorist associations and given his determination to exact revenge on those who oppose him, should we take the risk that he will not someday use these weapons at a time and the place and in the manner of his choosing at a time when the world is in a much weaker position to respond?
>
> The United States will not and cannot run that risk to the American people. Leaving Saddam Hussein in possession of weapons of mass destruction for a few more months or years is not an option, not in a post-September 11th world.

The Security Council didn't find Powell's evidence as compelling as did his president and colleagues at the White House. Powell presented photos of where the weapons *might* be stored and how they *might* be moved. He claimed there were U.S. intelligence reports of intimate connections between Saddam Hussein and al-Qaeda. We now know that there were no weapons of mass destruction and no intimate ties between Saddam Hussein and al-Qaeda. Iraq had nothing to do with 9/11, yet the groupthink that led to the disastrous decision to invade Iraq focused on 9/11 as the premise that trumped any argument to the contrary.

I know there are some who think it was a good thing to invade Iraq and that ultimately we ridded that country of an evil dictator, which makes it all worthwhile. One could argue that despite all the death and destruction the U. S. has caused, the evil done has been less than the evil perpetrated by the Saddam regime over many years. Following such a utilitarian principle, however, would lead to Armageddon. To justify a preemptive attack on a sovereign nation requires more substance than the desire to rid the invaded land of evil. In any case, the stated justification for invading Iraq was to protect the interests of the U. S, not to rid Iraq of evil.

Estimates vary, but the death toll of the Iraq war is between 100,000 and 1,000,000 Iraqis, most of them civilians, and more than 4,000 American soldiers. About 1,000 American allies have also died in the fighting, as have numerous journalists, media workers, and civilian contractors and mercenaries. The number wounded or maimed is in the tens of thousands. The destruction to property has been in the billions and the cost of living constantly under the threat of terrorist bombings in Iraq is incalculable. Such figures only partially account for the disastrousness of the decision to invade Iraq, a country that was no direct or indirect threat to the security of the United States of America.

At the time of the 9/11 attacks, Richard A. Clarke was the chief counter-terrorism adviser on the U.S. National Security Council. He was a holdover from the Clinton administration. Clarke claims that as early as the day after the attacks, Secretary of Defense Donald Rumsfeld was pushing for retaliatory strikes on Iraq, even though at the time al-Qaeda was based in Afghanistan. Clarke told Leslie Stahl of CBS's *60 Minutes* (March 24, 2004) that at a meeting with terrorism experts from the C.I.A. and F.B.I.:

> Rumsfeld was saying that we needed to bomb Iraq. And we all said ... no, no. Al-Qaeda is in Afghanistan. We need to bomb Afghanistan. And Rumsfeld said there aren't any good targets in Afghanistan. And there are lots of good targets in Iraq.... I thought he was joking.

According to Clarke, he and C.I.A. Director George Tenet told Rumsfeld, Colin Powell, and Attorney General John Ashcroft that "we've looked at this issue for years. For years we've looked and there's just no connection."

That was not what the White House wanted to hear. The National Security Council, as well as F.B.I. and C.I.A. experts, signed off on a statement that said there was no Iraq connection to 9/11 and sent it to the president. According to Clarke, it got bounced by the National Security Advisor or Deputy, who sent it back, saying, "Wrong answer. Do it again." Clarke told Stahl:

> I have no idea, to this day, if the president saw it, because after we did it again, it came to the same conclusion. And frankly, I don't think the people around the president show him memos like that. I don't think he sees memos that he doesn't—wouldn't like the answer.

Why not? Why wouldn't George W. Bush or Dick Cheney want to know the truth? I'd rather not speculate about their motives, but the facts indicate that there was a concerted effort by the Bush administration to provide a strong case for invading Iraq, which led men like George W. Bush, Dick Cheney, Colin Powell, Donald Rumsfeld, and others to systematically overestimate the worth of faulty intelligence reports and to underestimate the value of sound intelligence as they built their case to invade Iraq.

On or about February 9, 2007, a report by the Pentagon Inspector General claimed that Defense Secretary Donald Rumsfeld and his deputy, Paul Wolfowitz, had set up Douglas Feith as the leader of a Defense Department team that would provide "alternative" intelligence to the Bush administration. Feith provided intelligence that claimed good evidence for a Saddam Hussein/al-Qaeda connection. But the intelligence was intended to support the White House's political goal of invading Iraq rather than accurately reflect the conclusions of the U.S. intelligence community.

The White House didn't depend exclusively on the bad intelligence from its plant in the Defense Department. The Pentagon's Office of Special Plans, led by Abram Shulsky, also provided false claims about Iraq's connections to al-Qaeda. This Office was conceived by Paul Wolfowitz and began its work soon after the 9/11 terrorist attacks. It derived much of its intelligence from the Iraqi National Congress, an exile group headed by Ahmad

Chalabi, who wanted to overthrow Saddam Hussein. Chalabi apparently saw himself as eventually having a position of power and authority in a new Iraq. He now (October 2011) lives on a farm outside Baghdad and is a member of Parliament in the Iraqi National Alliance.

In his State of the Union address a few weeks before he sent Powell to address the United Nations, President Bush claimed that "The British government has learned that Saddam Hussein recently sought significant quantities of uranium from Africa." We now know not only that this claim was false but that Vice President Cheney urged his chief of staff and others within the Bush administration to leak to the press the identification of C.I.A. agent Valerie Plame as retaliation for an article ("What I Didn't Find in Africa") critical of the African uranium claim written by her husband, Joseph Wilson. Wilson had served as a diplomat in the U. S. Foreign Service from 1976-1998. He was assigned to the U. S. embassy in Baghdad when Saddam invaded Kuwait in 1990. Saddam threatened to execute anyone who protected foreigners in Iraq. Wilson sheltered many Americans at the embassy and successfully evacuated several thousand people. President George Herbert Walker Bush called Wilson "a true American hero." The action taken against Wilson's wife by the younger Bush's administration seems to support the cynical adage that no good deed shall go unpunished.

There is no decision more important that a President of the United States can make than the decision to send soldiers into battle, except perhaps the decision whether to use nuclear, chemical, or biological weapons during a war. The moral requirement to hear *all* the evidence and to carefully evaluate it before making a decision of such magnitude cannot be higher. To push away dissenters and draw closer those who agree with your gut feelings is the worst thing an executive can do when making life and death decisions. It is especially important at such times that dissenting opinions be heard and not summarily dismissed. The result of the Bush administration's decision has been thousands of maimed and dead men and women from the United States and the countries of its allies, tens of thousands of dead Iraqis, and millions of lives ruined as people try to live on with injuries, without loved ones, in a devastated country or on the run in foreign lands. A nation that was once the cradle of civilization is now a bombed-out monument to what bad decisions by world leaders can produce. This fact seems to support another cynical adage: the road to Hell is paved with good intentions.

Some leaders learn from their mistakes. Others plod on, oblivious to criticism and sure of their decisions despite what the rest of the world might think. Before Donald Rumsfeld resigned as Secretary of Defense, he set up an "Iranian directorate" in the Pentagon headed by none other than Abram Shulsky. Thus, when the Bush administration started claiming that it had good intelligence that Iran was endangering U.S. forces in Iraq by supplying Shias in Iraq with weapons and electronic devices for detonating bombs it was not surprising that the claims were met with skepticism. It seems obvious to some people that there can be only one reason to put Shulsky in charge of gathering intelligence on Iran: he's there to provide "alternative intelligence" that could be used to justify Bush in attacking Iran. Whether the skeptics were right or not about the desire of the Bush administration to make war on Iran, its past behavior justified the mistrust of its intelligence on Iran.

Skepticism regarding U.S. intelligence on Iran continues under the Obama administration. Reports of an Iranian plot to assassinate Saudi Arabia's ambassador to the United States and of Iran's imminent possession of nuclear weapons have been met with mixed reactions. Some scoff at the reports as disinformation setting the stage for an invasion of Iran. Others take the intelligence reports at face value and conclude that Iran must be stopped. The average citizen is justified in being skeptical of government-issued warnings based on intelligence reports. We have no way of knowing what the truth is when we hear such reports.

Groupthink in Action

Ford's decision in 1958 to create the grotesque-looking Edsel, named after Edsel Ford, and President John F. Kennedy's decision to invade Cuba in 1961 are often cited as examples of disastrous decisions due to groupthink. Psychologist Irving Janis defined groupthink as "a way of deliberating that group members use when their desire for unanimity overrides their motivation to assess all available plans of action." Another example of groupthink that is often cited is the *Challenger* Space Shuttle disaster, where Thiokol executives became insulated, conducted private conversations under high stress, and were afraid of losing potential future revenue should they disagree with NASA. Thiokol made the O-rings that failed in the cold conditions causing the solid rocket booster to leak fuel and explode, killing seven crew members. NASA officials ignored warnings that contradicted the group's goal, which was to get the launch off as soon as possible.

Ford's Metallic Monster: the Edsel

Some of the symptoms of groupthink that lead to bad decisions are also symptoms of techniques used by individuals who make bad decisions. If you think you're invulnerable, you're asking for trouble. If you rationalize or gloss over criticisms, you are asking for trouble. You are likewise asking for trouble if you don't accept the possibility that you might be wrong or if you label and stereotype critics as unworthy of consideration or do anything else to prevent contrary viewpoints from being heard.

What does a great leader do when faced with gathering information and planning for the future? Jim Collins argues in *From Good to Great* that a great leader gathers the best people he or she can find and gets their input *before* making plans or decisions that are going to affect the company, its employees, and the stockholders. Unfortunately, too many leaders in both politics and business surround themselves with people who think alike and who believe that they'll advance more quickly and further if they don't rock the boat. When newly elected as governor of California, Gray Davis told the media that the legislature's job was to implement his vision. He made it clear that he was not a great leader. In 2003 Davis became the second U. S. governor to be recalled. He didn't have too many friends in the state legislature defending him.

One result of surrounding oneself with sycophants is that not all alternatives are considered; the only options that get considered are those that are seen as promoting what the group members think the leader wants. Group members tend not to offer ideas that might be seen as critical of the leader. On the other hand, group members are quick to attack anyone whose ideas conflict with the group's mindset. A defensive wall is built around the mindset; all criticism is seen as obstruction and must be defended against at all costs. Furthermore, the group tends to ignore expert help unless that help supports its mindset. This selectivity in how the leader or the group gathers and evaluates information can lead to disastrous consequences. Furthermore, because the group members keep reinforcing each other and pushing out those who disagree, the group becomes overconfident in its decisions and often fails to create contingency plans in case their designs fail. They don't allow themselves to consider failure as an option. So, when they fail, they fail in grand fashion—unless of course the government bails them out or they *are* the government and they can hire the best propagandists money can buy to make it look like their failures are actually successes.

Great leaders, on the other hand, make contingency plans and encourage a variety of opinions and options. Great leaders don't gather people around them for the purpose of implementing the leader's vision. Great leaders use the experience and knowledge of the people around them to create that vision and to figure out how to best implement it. Groupthink plagues leaders who are not so great. Some of the traits of groupthink are:

1. Excessive optimism that encourages taking extreme risks;
2. Excessive confidence that excludes reconsidering assumptions;

3. Excessive demand for conformity that leads to self-censorship and to seeing critics as enemies.

Great leaders, on the other hand, take risks that are based on rational and fair-minded assessments of *all* the relevant data. They encourage their people to express their feelings and beliefs, but require them to back up their claims with data. They are not afraid to examine their assumptions. They see critics as friends rather than as enemies. Critics force you to examine your decisions, your assumptions, and your values; they don't allow you to get away with sloppy thinking. Hence, they are a valuable asset.

It is not an easy task to create an atmosphere where members of a group feel confident and free to explore ideas without fear of being isolated or excluded if their ideas are rejected or deemed unsuitable. The natural tendency is to require conformity; too much freedom is seen as a threat to order that can lead to chaos. Or worse, it could be seen as a sign of weakness in your leadership ability. It is difficult to be truly open to *all* ideas that your group members bring to the table. Decisions must be made and those whose ideas are followed will have a tendency to see themselves as *winners* and those whose ideas are rejected may see themselves as *losers*. Maintaining equanimity and keeping all the group members happy, especially when their suggestions or plans are not accepted, is not easy. A great leader provides good reasons for choosing one plan of action over another. If a leader shows favoritism or rejects ideas for obviously inadequate reasons, resentment will follow. You don't want people in your group who feel resentment or sense favoritism when their ideas are rejected *for good reasons*. You want team players but you don't want blind followers. Good luck!

Most of my experience with groups has been as a coach and as a faculty member on college committees. As a coach, I didn't allow group decisions or even ask for input from my players. I suppose you could say that I acted like a dictator. Such leadership seems to be fine for many kinds of athletic teams, but it would be inappropriate for corporate, educational, or political leadership. Unfortunately, being inappropriate hasn't hindered the Coach from being the leadership model for many CEOs and presidents. The Coach is like the drill sergeant or military commander who wants followers, people who will take orders, carry out tasks, and not question decisions. The metaphors of the Coach and the Team Player are misplaced when the desired outcome is good decision-making for the company, the school, or the country.

I've been on both effective and ineffective committees. The effective ones were led by managers who welcomed ideas and encouraged members to speak up and offer critical comments on any proposal before the group. The successful leaders often used brainstorming techniques, where committee members would spend a set amount of time just trying to get ideas onto the table or whiteboard. In brainstorming, you don't criticize ideas as they're proposed. Criticism too early in the decision-making process can stifle creativity by intimidating people. Once all the ideas are collected, they're organized and presented to the group for consideration and critical evaluation. Brainstorming seemed to work well enough, but I wasn't aware of other options at the time. Since retiring I have learned that this method has been shown not to work as well as having the members come to the table with their independently derived thoughts. In any case, ultimately the administrator has to make the decision and take responsibility for it, but the committee members need to feel that their ideas were taken seriously even if they were ultimately rejected. The worst leadership came from those who seemed to have their minds made up before discussion of a problem or issue even began. Such managers were not truly interested in our ideas; they wanted us to rubber stamp their own ideas. Often, their ideas weren't even their ideas; they were their boss's ideas. Hierarchies of managers pose their own peculiar kinds of problems for an organization that wants to encourage critical thinking. Too much management is top down rather than bottom up. Those at the bottom, who are in the trenches, so to speak, are often a manager's best but least used assets. I once worked for a dean of instruction who had never spent a day in the classroom as a teacher, but she was the best dean I ever worked for because she consulted and listened to teachers before making decisions. I worked for a couple of other deans who were career teachers before moving into management but who were mediocre managers at best because they waited for someone above them in the hierarchy to give them direction. They consulted with teachers only *after* the negative consequences of their actions became apparent. Even then, the teachers had to initiate the engagement because the managers wouldn't face up to their errors unless forced to. I'll give you a trivial example.

At my college, any request by a teacher for time off to go to an out-of-state conference (even as a speaker or workshop coordinator) requires a written request at least one month in advance from the teacher's dean to the chancellor of the college district. I wanted several days off to participate in a workshop on critical thinking that I had coordinated for the James Randi Educational Foundation's "Amazing Meeting," which is held in Las Vegas, Nevada. My college is in California. I got my paperwork in on time, but my dean failed to send the written request in at all. He probably didn't know it was required, though he should have known it. Anyway, when the travel paperwork came back to me after going through various college channels, I was informed of this oversight. I was also informed that the dean of instruction (not the one I praised above!) had decided that the best thing to do (for her, I'm sure) was to have me take sick leave for the days I took off. I objected that I hadn't taken sick leave and her decision didn't make any sense. I was able to get that decision rescinded by our district chancellor. What kind of manager would make such a decision? One who should have notified my dean of the required request letter to the chancellor's office and who didn't want anyone to know that she had not done so. She could also avoid some paperwork. Needless to say, such management decisions do not inspire trust or employee loyalty, two qualities essential to good management.

The worst of the worst leaders, however, are those who reject somebody's ideas and then later introduce them to the group as their own and those who don't involve in their decision-making those who will be profoundly impacted by their decisions. Anyone who has had the misfortune of being a member of a group with an inept leader can supply plenty of examples of leaders rejecting ideas and then later putting them forth as their own. I'll supply an example of a decision by a leader who failed to consult those who would be most affected by his decision. One day I arrived on campus to find that all the automatic doors for the handicapped had been changed. If you were in a wheelchair, you had to wheel up close to the door and hit a large metal plate to open the door. The door then opened into you as you tried to enter or exit a building. You had to be very quick and agile to get out of the way of the door as it opened. Of course the handicapped could not use these doors without endangering their safety. The doors had to be removed and replaced with a new type of automatic door. The administrator who ordered these doors forgot the first two rules of intelligent decision making: (1) When you make a change that affects others, walk in their shoes first; and (2) consult, if possible, with those most affected by the change. The one who ordered these doors should have consulted with a group of handicapped people before making his decision and he should have put himself in a wheelchair and tried out a simulation of the new procedure before implementing the change. Consulting with those who will be most affected by a decision can prevent much grief and might even save the organization some money by not having to redo something that wasn't thought through carefully enough. I've seen a registrar institute registration changes for students that he never would have instituted had he forced himself to act as a student and try to follow his recommended changes, which involved standing outside in line for hours at a time. These rules should be at the top of the list for anyone making decisions affecting people whose cultural traditions differ from the decision maker's. Offense may be given where none was intended. A simple consultation with those who would be affected could prevent even acts of kindness from turning into perceived insults. For example, when U.S. military officials decided to give presents to Afghan children, they should have consulted with some Islamic leaders before having soccer balls made that were inscribed with the *Shahada*, a declaration that there is no god but Allah and that Muhammad is his messenger. Seeing the names of Allah and Muhammad on an object intended to be kicked around in the dirt offended many Muslims. Instead of being thanked for their generosity, the U.S. was criticized for its insensitivity.

One final word on committees: college committees are usually staffed with volunteers; they're not usually handpicked by the administrator in charge. Nothing makes work on a committee less bearable than an obnoxious colleague who rides a one-trick pony. Leaders who do not know how to shut up disruptive members will find their effectiveness waning in proportion to the volume and intensity of the Johnny-One-Notes with their pet peeves, special interests, or self-proclaimed special expertise.

While I am not a businessman and have never owned or run a business, I sometimes get asked by business persons why management doesn't think critically. I assume they ask this question because of some personal experience from which they have generalized. I tell them that I have had experience with the management of only one college. I'm sure many organizations support critical thinking, but those that don't will share in common certain features. If the managers of an organization don't think critically it may not be because they are untrained or incapable of critical thinking. It may be because they are not rewarded for it. In fact, they're often punished for it. Many managers surround themselves with yes-people, but thinking critically requires looking at all the options, not just the ones that will please the boss. Critical thinking requires freedom to think and question. Many organizations don't want people who will think and question; they want people who will "do it my way or it's the highway." Creating an atmosphere where managers are not afraid to express themselves or explore ideas may be difficult. Such an atmosphere requires trust. Getting people to trust you when you say you want them to let you know when they think something could be improved may be easier said than done. Many students have had teachers who encouraged them to ask questions, but when they did they found that instead of being treated with respect they were made to feel stupid, especially if the question touched some nerve that the teacher was especially sensitive to, like a political opinion. Employees who want to climb the corporate ladder figure out what kind of behavior is rewarded. If they see that the critical thinkers never advance or are sent to "outer Siberia," they will keep their ideas to themselves. Of course, having lots of ideas is not enough to bring success to any organization. As two-time Nobel Prize winner Linus Pauling once said when asked how he came up with so many great ideas: *the trick is to have lots of ideas and then throw away the bad ones.* (Unfortunately, Pauling didn't follow his own advice regarding the health benefits of vitamin C.) It takes critical thinking to evaluate ideas and distinguish between the good and bad ones, but first you must get the ideas. Any manager who systematically blocks the creative flow of ideas will be limiting options. Giving people the freedom to think won't guarantee that they will think great thoughts, but stifling thought guarantees that many great ideas will be lost, perhaps forever.

Historically, there is probably no area of human existence that has been more negatively affected by communal reinforcement than religion, especially those religions with sacred texts and authorities who demand orthodoxy and unquestioning acceptance of the "divine word." Politics, however, comes in a close second. Much of the groupthink in both religion and politics is driven by fear. In politics, at the worst extreme there are leaders like Adolf Hitler, Joseph Stalin, Idi Amin, Saddam Hussein, Muammar Gaddafi, and Bashar Assad who threaten dissent with death. In religion, the worst leaders are those who have threatened with death and eternal damnation those who are not subservient to their will.

Fear works on another level as well. To abandon the faith means abandoning one's family and friends. It means withdrawal from the community that has nourished, raised, and protected you all your life. You will be rejected and humiliated. You will be labeled an infidel or a traitor if you dissent from the party line.

Ideas that no rational person today would derive by observation and reflection are accepted by millions of people because they are taught to them as children by adults who guide them and protect them in many ways. Belief in apparitions, demonic possession, incarnations of gods, karma, miracles, prophecies, reincarnation, teleportation, transubstantiation, the Holy Trinity, and witches are believed by billions on the basis of faith, i.e., irrational acceptance reinforced by the community of believers. Furthermore, as group leaders demand signs of loyalty and faith, lines are drawn that identify members of other political groups, nations, or religions as enemies to be feared and hated. The leader is not just to be trusted and admired, but revered. Questioning the main beliefs and policies of the group is considered treasonous or heretical and can get one labeled as an enemy. In some religions or countries, this can be fatal.

Communal reinforcement accounts for the popularity of many unsupported claims, e.g., that children have memories that are completely accurate, that children rarely says things that aren't true, that you can rid yourself of cancer by visualization or humor, that Jews control all the power and money of the world, that you can rid yourself of heart disease or become incredibly rich just by changing your attitude, and so on.

Communal reinforcement explains how entire nations can believe in such things as witchcraft or demonic possession. It explains how testimonials reinforced by other testimonials within the community of therapists, sociologists, psychologists, theologians, politicians, talk show aficionados, and so on, can be more powerful than scientific studies or accurate gathering of data by unbiased parties. In addition, communal reinforcement can be used to build shields against those who might challenge the status quo, something which many religious and political leaders have discovered in their efforts to neutralize criticism.

One way to control criticism is to manage the information that people are allowed to have access to. One need not be a communist leader with a state-controlled press to be in command of information flow. Democratically elected leaders are very gifted at managing the news and controlling the flow of information and misinformation, as we shall see in the following chapter.

SOURCES FOR CHAPTER FIVE 272

VI
Reliable Sources of Confusion, Collusion, and Spam

"No people can be really free if its press is spoon fed with government pap or if the news which provides a democracy with the rationale for its actions is so controlled, restricted, managed, or censored that it cannot be published." —Hanson W. Baldwin

---

We're constantly bombarded with 'information' from mass media news providers, politicians, scholars, scientists, talk radio hosts, television pundits, government agents, and so on. Sifting out reliable and useful information from all the sexy garbage and propaganda thrown our way is becoming nearly impossible for the average person. This chapter will explore ways to evaluate sources of information.

Adding to the difficulty of identifying reliable sources is the problem of covert propaganda: systematic, but below the radar, disinformation from government sources, corporation shills, bloggers, and religious ministries. Governments and corporations hire people to pose as journalists or to write letters to editors as if there were a grassroots movement swelling up in favor of some politician or corporate executive. Phony news stories and emails are sent out, hoping to affect the stock market. It's hard to believe, but apparently many people make investments or base health decisions on the words of total strangers who send them unsolicited emails or, as it is otherwise known, spam.

Encouraging Supportive News Coverage

Vice President Dick Cheney complained in an interview with conservative media commentator Armstrong Williams about bias in the press. Cheney was absolutely right about bias in the press. The Bush administration had paid Williams $240,000 to promote the No Child Left Behind Act in broadcasts and in print. Williams lost his job with Tribune Media Services when it was revealed that he was a hired propagandist producing covert puff pieces for the Bush administration. Cheney didn't lose his job, even though using government money for propaganda purposes is illegal. The paper trail didn't lead to his door but to the door of the Education Department. "They should fire everyone who was involved in this," said Rep. George Miller, D-Calif. at the time. "This was not an accident, this was not an oversight. This was an intentional effort to corrupt the process." As far as I know, nobody got fired, fined, or even reprimanded, chided, or scolded for using government money for propaganda purposes.

This was not the first time Cheney had manipulated the press while pretending to be an innocent. The Bush administration duped *New York Times* reporter Judith Miller about Saddam Hussein's weapons of mass destruction. "Reliable sources" provided Miller with information about aluminum tubes that were allegedly being acquired by Saddam for processing uranium. The information was bogus, but when Cheney appeared on NBC's *Meet the Press* with Tim Russert he cited the *New York Times* article to bolster the Bush administration's case about weapons of mass destruction. The story had been on page one that morning. Such opportune coincidences occurred with some regularity during the Bush/Cheney years. "You leak a story, and then you quote the story," noted CBS news correspondent Bob Simon in an interview with Bill Moyers. Yes, and the rest of us don't find out about it until it's too late. Russert, by the way, told Moyers he had no idea he was being manipulated by Cheney.

Another way to control the response to stories that are going to find their way into the mass media anyway is to release the bad news late on Friday or on Saturday. Cathie Martin, former Cheney communications director who testified in the perjury trial of former Cheney chief of staff I. Lewis "Scooter" Libby, said: "Fewer people pay attention to it late on Friday...[or]...when it's reported on Saturday." (In March 2007, Libby was found guilty of four felony counts of making false statements to the F.B.I., lying to a grand jury, and obstructing an investigation into the leak of the identity of undercover C.I.A. officer Valerie Plame.)

Another way the Bush administration tried to manage the news was by giving press credentials to fake journalists so they could lob softball questions to the president at press conferences. Their star performer was a fellow named James Dale Guckert, who worked under the pseudonym Jeff Gannon. Guckert represented a dubious news agency called Talon News as its White House reporter between 2003 and 2005. The Bush administration used him as an agent of disinformation. The Justice Department questioned Guckert about his mentioning of an "internal government memo prepared by U.S. intelligence personnel" that implicated Valerie Plame as the one who suggested that her husband investigate the claim that Iraq was attempting to procure yellowcake uranium from Nigeria. Guckert is unrepentant. He ran an anti-Democrat website on which he posted self-serving screeds like the following:

> The discredited former C.I.A. analyst and her blowhard husband have moved to New Mexico, leaving behind their legacy of deceit. The pair have [sic] relocated to Sante [sic] Fe, far away from the scene of their crimes. I generally refrain from such harsh language, but these two have perpetrated an enormous fraud against the United States and undermined the global war on terror. How many American citizens will die because Plame and her ilk leaked classified information about effective counter-terrorism programs to *The New York Times* and *Washington Post*? (www.jeffgannon.com; accessed 5/5/07)

Guckert's Jeff Gannon website has vanished, but Guckert has risen from the ashes and in 2008 he became a member of the National Press Club in Washington, D.C. Was his website blog underwritten by conservative Republicans with an agenda that included protecting the vice president and his chief of staff? Who knows? What we do know is that many bloggers and websites that appear to be independent are actually fronts for special interests in the government or the private sector. If you get your news from bloggers and think you're getting a fresh, independent viewpoint, you may be deceiving yourself. Anyone can set up a website and make it look like anything they want. There is a lot of good information available on the Internet, but you often have to run through fields of land mines to get to it.

You may remember a swell-sounding outfit called the Global Climate Coalition. It was founded around 1990 and funded by such independent folks as Exxon, Ford, Texaco, General Motors, British Petroleum, and DaimlerChrysler. The purpose of the coalition was to fund scientists and writers who would provide disinformation regarding climate change to anybody who would listen. For more than ten years they supported "research" that would cast doubt on scientific reports warning of the impending dangers from increased greenhouse gases caused mainly by the burning of fossil fuels. They didn't care whether the "research" was any good; what mattered was that it raise some doubts regarding the scientific consensus and produce an "opposing viewpoint" that fair-minded journalists are supposed to give equal time to. After Bush and Cheney took power, there was no further need for the Global Climate Coalition. It was "deactivated" in 2002.

Even though the Bush administration did little to encourage the proper study of climate change, they tried to appear to do so. On February 7, 2007, White House Press Secretary Tony Snow boasted that the United States surpassed Europe in reducing greenhouse gas emissions. When challenged to prove it, Bush's science advisor John H. Marburger announced:

> According to the International Energy Agency, from 2000-2004, as our population increased and our economy grew by nearly 10%, U.S. carbon dioxide emissions increased by only 1.7%. During the same period, European Union carbon dioxide emissions grew by 5%, with lower economic growth.

What Marburger failed to note was that the drop during this period was due to decreased air travel because of a modest recession and fear after 9/11. Marburger cherry-picked the data, much like some mutual fund salesmen do. Use any year but 2000 as your starting point and the European Union surpasses the U.S. in performance. (The graph that the mutual fund salesman shows you always pick a starting point that has you making a fortune had you invested at that starting point. Pick another starting point and you can show how much money you would have *lost* had you invested then.)

Another way the Bush administration tried to manage the news was by distributing prepackaged, ready-to-serve news reports to TV stations, a tactic long-used by corporations and special interest groups. The State Department, for example, produced a video news release (VNR) featuring a euphoric Iraqi-American in Kansas City thanking Bush and the U.S.A. for invading Iraq. Another VNR, produced by the Transportation Security Administration, featured a fake reporter covering airport safety. The reporter was actually a public relations fellow who reported on "another success" in the Bush administration's "drive to strengthen aviation security." The fake reporter called it "one of the most remarkable campaigns in aviation history." The Agriculture Department's office of communications also produced a propaganda piece for the Bush administration about its determination to open markets for American farmers. According to the *New York Times*, at least twenty federal agencies produced VNR propaganda pieces under the Bush administration, including the Defense Department and the Census Bureau (May 13, 2005). None of the fake reporters used in the VNRs mention that they work for the government and each VNR clearly supports one of the administration's policy objectives.

Several Bush administration VNRs featured a fake reporter who ended her reports with "In Washington, I'm Karen Ryan reporting." Ryan is a public relations person who narrated a video praising Bush's program for remedial instruction and tutoring of children. It "gets an A-plus," she intoned. The video was made by Home Front Communications. Ryan also did two VNRs praising Bush's policy on Medicare drug benefits. The Government Accountability Office labeled the programs "covert propaganda" because the VNRs don't identify who made them.

Media outlets were happy to oblige the Bush administration in its efforts to promote its policies via fake news reports. Television stations get professional looking news reports at no cost—except to their credibility when the public discovers it's been deceived. There is no way the average viewer can know that what she is watching is not the result of an independent investigation unless the maker of the VNR is identified. Rarely does a television station identify who made the VNR. When the station runs its logo over the showing of the VNR—as CBS did for a VNR produced by some trial lawyers on seat belt hazards—the station deceives the viewer into thinking the network investigated the story. The

Clinton administration also produced VNRs, but the Bush people turned it into an art form. (The Obama administration may not get the opportunity to promote any of its programs because Republicans seem to be intent on blocking any program his administration puts forth.) Producing your own news heads off critics and provides a powerful propaganda tool to counterbalance the effect of real news done by independent agencies. For example, on September 11, 2002, WHBQ, a FOX affiliate in Memphis, Tennessee, showed a very positive piece on how the United States was liberating the women of Afghanistan:

> Tish Clark [now Tish Clark Dunning], a reporter for WHBQ, described how Afghan women, once barred from schools and jobs, were at last emerging from their burkas, taking up jobs as seamstresses and bakers, sending daughters off to new schools, receiving decent medical care for the first time and even participating in a fledgling democracy. Her segment included an interview with an Afghan teacher who recounted how the Taliban only allowed boys to attend school. An Afghan doctor described how the Taliban refused to let male physicians treat women. (Barstow and Stein: 2005)

According to Clark, the U.S. was liberating women and improving their lives. Clark didn't mention that State Department contractors, not people in FOX's news division, made and edited the film and the interviews her station used. Clark's narration was mostly read from a government-supplied script, though she claims she didn't know that the report originated with the State Department.

According to David Barstow and Robin Stein of *The New York Times*, President Bush's communications advisers devised a strategy after Sept. 11, 2001, to "encourage *supportive news coverage* of the fight against terrorism" (italics added). That's one way to describe fake news.

State agencies also have jumped on the VNR bandwagon. The Texas Parks and Wildlife Department produced about 500 VNRs between 1993 and 2005. Three labor unions sued California governor Arnold Schwarzenegger for making VNRs that violate the state's law against using government money to produce propaganda promoting a policy position. Sacramento Superior Court judge Lloyd Connelly ruled that the Schwarzenegger administration had undermined "the public's ability to participate in the rule-making process." By including only supporting comments from the public or proposed regulation, wrote Connelly, the fake TV news reports created "the misleading impression that the regulations are unopposed by any segments of the public and are not subject to criticism, thereby discouraging any further questioning or investigation of the matter by the public." The proposed regulation had to do with such things as lunch break guarantees and mandated nurse-to-patient staffing ratios.

Fake news can be used *against* those in power, too. Millions were spent by people like Richard Mellon Scaife and Patrick Matrisciana to hire "journalists" to dig up dirt on Bill Clinton and publish it (Carroll 2005: 67). Politicians and their enemies, however, aren't the only ones who try to manipulate the media. The police, district attorneys, generals, corporation leaders, and anyone who has an interest in manipulating public opinion is a likely candidate for someone who will try to control the content of information that passes for news these days. When Microsoft was being pursued by the Justice Department for antitrust activities, it created a number of front groups and poured money into several think tanks and other organizations willing to espouse Microsoft's views in publications. It also instigated phony grassroots letter writing campaigns with a pay-as-you-write policy. According to the *Wall Street Journal* (10/20/2000), a pro-Microsoft letter to a member of Congress from a mayor or local Republican Party official was worth $200. A letter or visit by a fundraiser known to the lawmaker or a family member could be worth up to $450. An op-ed piece in local papers could fetch $500.

Failing Journalism 101

Fake stories are always deceptive, but they are not always malicious or manipulative. In 1983, *Sacramento Bee* reporter Dan Stanley covered a lecture given at the University of California at Davis by feminist leader Gloria Steinem. His report was published, but many of those who had attended the lecture and read Stanley's account of it wondered why he reported things that Steinem didn't say. They also wondered why he didn't mention the bomb threat. He didn't mention the bomb threat because he had left the auditorium before the lecture began and didn't know about the bomb threat. He reported things that Steinem didn't say because he interviewed her backstage and asked her what she was going to talk about. Since he didn't stay to hear the lecture, he had no way of knowing that her talk didn't match the interview. Stanley had a good reason for not staying to hear the talk—he had a deadline to meet—but he didn't mention that reason in his report. Instead of mentioning that the auditorium had been cleared and the talk rescheduled to the point where he would miss his deadline if he stayed, he wrote his article as if he were reporting on what he heard Steinem say to the audience. As journalistic transgressions go, this one is trifling and hardly worth mentioning when compared to the lies and deceptions of other journalists. Stanley was reprimanded. Others who faked their stories for the *Bee* were not so lucky. Sportswriter Jim Van Vliet was fired after writing a story about a San Francisco Giants baseball game that he watched on television in a bar. He wrote the story as if he had interviewed players at the stadium. Diana Griego Erwin, a Pulitzer Prize-winning columnist, quit after a fact-checker found that several of the people she allegedly interviewed and wrote stories about didn't actually exist.

Fake stories have sometimes been produced by journalists just because the fake story seemed like a better story than the unvarnished truth. Janet Cook won a Pulitzer Prize for her story about an 8-year-old heroin addict named "Jimmy." She was fired and gave up the prize when it was discovered she didn't really attend Vassar College, as her resume indicated, and she confessed that "Jimmy" did not exist. Stephen Glass was fired by *The New Republic* magazine for fabricating many stories, including one featuring a 15-year-old computer hacker who broke into a large company's computer system and was then offered a job by the company. Patricia Smith resigned from the *Boston Globe* after it was discovered that she was using fictional people in her allegedly factual columns. Mike Barnicle also resigned from the *Boston Globe* amid charges of fabrication and plagiarism. *New York Times* reporter Jayson Blair resigned after it was learned that he fabricated interviews and other important details in at least 36 articles. Jack Kelley of *USA Today* resigned after his bosses found letters on his computer that he'd written to friends, asking them to pretend to be sources when editors verifying his stories called them. Kelley had been fabricating stories for more than a decade. On April 12, 2007, we learned that "Katie Couric's Notebook" columns on the CBS News website aren't written by Katie Couric, even though the column is written as a first-person account of supposedly real events. We found out about this deception when CBS announced that it had fired an unnamed person for being responsible for posting a plagiarized article under Katie's name.

The above transgressions may represent the tip of the proverbial iceberg or they may be aberrations. These are just a few high profile examples of lying and deception by journalists. Can we trust journalists, then, even though some of what they present is faked, plagiarized, or handed to them by interested parties? Is the information we are getting on a daily basis generally reliable? You'll have to decide that for yourself, but remember that most of the information I have used to describe attempts at manipulation of and by the media has come from journalists themselves.

Even when journalists try their best to get the story right, however, they can unintentionally manipulate us into believing what isn't true. The examples I use here have all been widely reported in the press. The reader should have little difficulty in seeing how these examples demonstrate the dangerous consequences of stereotyping of the mentally ill, profiling of the criminal mind, and racism. Preconceived ideas about the mentally ill, the terrorist, the dangerous black man, and the "privileged white man" can affect how the news is reported and how the public receives that news. These stories also show us how the police can manipulate the press.

Law Enforcement and the Media: A Dangerous Liaison

The website for the city of Gainesville, Florida, declares that it sets the standard for excellence for a top ten mid-sized American city. The university town of about 100,000 was sent into a panic on

August 26, 1990, however, when the news media reported that two young women had been mutilated and murdered in their apartments. The next day another young woman's body was found. She too had been mutilated and murdered. The following day two more bodies were found, a young man's and a young woman's. All the dead had been college students. The top-ten town went into panic mode. The news media reported the grisly events in graphic detail. Fear gripped Gainesville's citizens as well as thousands of parents whose sons and daughters were students at the University of Florida. Within a week, the townsfolk could rest easy again, however. The media reported that the police had arrested their prime suspect in the gruesome killings. A University of Florida freshman, 18-year-old Edward Lewis Humphrey, who was described as unbalanced and violent, had been taken into custody. Earlier that summer, Humphrey had been kicked out of the apartment complex where the last two victims had been found. His roommates said he was "weird." It was reported that he had thrown a chair at the apartment manager when he tried to get Humphrey to leave the premises.

Humphrey had been in trouble with the law before and had a history of bizarre behavior, like walking into people's apartments uninvited and peeping through their curtains when they locked him out. There were also several reports of him behaving violently toward different people, including his grandmother as she took him around town in search of a place to live. The media reported that Humphrey had been diagnosed with manic depression—also known as bipolar disorder—and had exhibited strange behavior after he had stopped taking his medication. There was also a rumor that he had been romantically interested in one of the victims. The news media reported that he dressed in military-type fatigues and went on late-night "reconnaissance missions" carrying a hunting knife. Police put him under surveillance. He was arrested after his mother called police and reported that he had hit his grandmother during an argument. He was taken to a medical center. Later, the F.B.I. interrogated him.

Humphrey was never charged with the Gainesville murders, however, because there was no evidence linking him to the crimes. He was charged with battery of his grandmother, even though she declined to press charges. The police did not let the public defender meet with Humphrey, but they held him, apparently hoping to find evidence that he had perpetrated the brutal killings. Bail was set at $1,000,000, which is a bit steep for someone accused of hitting a grandmother who declines to press charges against her grandson.

The media made it clear that the police had their man, declaring Humphrey the "Number One Suspect." The police actually had several other suspects, but the only name they released to the media was Humphrey's. His mug shot appeared in the papers and on television, but, as already noted, he was never charged with the murders. The million dollar bail indicated that somebody in power thought he was the killer or at least wanted the public to think he was the killer. The media apparently didn't think it was odd that it took the police several days before they could persuade a judge to let them search Humphrey's apartment and his grandmother's house. The searches yielded nothing linking Humphrey to the murders, but they did manage to frighten his grandmother to the point where an ambulance had to be called.

The fact that there were no more similar murders after his arrest fed the media's conviction that the police had the right man. Every day from August 30 (the day he was arrested) through September 11 Humphrey was front page news. The police and the media had the townsfolk calmed down by September 12. The killer was in jail and the citizens of Gainesville could feel safe again. In the end, Humphrey was tried and found guilty of battery—even though his grandmother testified that he never hit her. He was sentenced to 22 months in a state hospital. Their collective nightmare apparently over, Gainesville could return to its top-ten insouciance.

Four years after Humphrey had been tried and convicted by the press, Danny Rolling pled guilty to the Gainesville murders. He was put to death by lethal injection on October 25, 2006. Before Humphrey had been indicted, the media had camped out around his grandmother's house. She died of a heart attack while arguing with a reporter who had been pounding on her door, demanding that she answer questions. To Humphrey's credit, he graduated from the University of Central Florida in 2000 with a degree in business administration.

The media, with the aid of the police, had all but convicted Edward Humphrey, an innocent man. The journalists who covered the story might blame law enforcement for misleading them, but they were willingly led, perhaps because they too were convinced of Humphrey's guilt. As far as I know, neither anyone in law enforcement nor any journalist has ever apologized to Humphrey or his family. Of course, Humphrey was not the first nor will he be the last person unjustly accused of a crime by both police and media. Remember Richard Jewell, the security guard at the Atlanta Olympics in 1996 who discovered a pipe bomb minutes before it exploded. He was unjustly accused of planting the bomb. The *Atlanta Journal-Constitution* ran stories that implicated Jewel. They reported that the F.B.I. said Jewell fit their profile of the "lone bomber." They reported that Jewell was "an individual with a bizarre employment history and aberrant personality." Jewell sued for libel but the paper refused to settle. Tom Brokaw of NBC News said on the air: "…they [the F.B.I.] probably got enough to arrest him. They probably have got enough to try him." Those comments seem innocent enough when taken out of context, but given the national media blitz focusing on Jewel, they implied the guilt of a man who hadn't even been arrested. The F.B.I. was very aggressive and very public in its investigation of Jewell. The news coverage was so intense that two of the bombing victims filed lawsuits against Jewell. Jewel sued NBC and allegedly received more than $500,000 in a settlement. Eric Rudolph was charged with the Olympic bombing, along with some other bombings that he said he did as part of a campaign against abortion and "the homosexual agenda." Jewel went on to become a police officer. He was eventually honored by the governor of Georgia for his efforts before and after the Olympic bombing. U.S. attorney General Janet Reno almost apologized to him. "I'm very sorry it happened," she said in a press conference. "I think we owe him an apology." I think so too.

When a white man who has been shot in the stomach tells police he and his slain wife were shot by a black man in a jogging suit and the police have no reason to doubt him, they tell reporters what the man told them and the story goes to press. When a white woman cries hysterically that a black man stole her car with her two young children in it, the police listen and the press reports it. Later, we find out that the white man had murdered his wife and shot himself, and the white woman had sent her two children to their deaths by strapping them in her car and sending it into a lake. No one was arrested in the alleged carjacking except the white woman who killed her children, but William Bennett was arrested and all but convicted by the Boston police and press before Charles Stuart's brother went to the police and told them what had actually happened. Stereotypes and the prevalence of bias, especially racial stereotypes and racial bias, make is easy for people to manipulate the police and the press. Sometimes the police manipulate the press and sometimes the press manipulates everybody. How are we to know when the information we get from the police or the media is reliable? The task is not easy.

The first lesson to draw from the above anecdotes is that a critical reader must not assume that stories we read in the press or hear on TV are true, even when the source of those stories is the F.B.I., the district attorney, or some local law enforcement agency. This sounds obvious, but it is more difficult than you might think to withhold judgment on someone who has been pilloried in the press by a one-sided presentation of the evidence. The combination of television, newspaper, magazine, radio, and Internet reporting can be overwhelming. Stories that have little merit pick up added energy from the reinforcement effect of so many different sources repeating the same message. They give the illusion of getting the same story from several different sources, when often they are all getting their information from a single source who may be biased or making stuff up. The task of the critical reader is similar to that of the fair-minded juror: don't pass judgment until all the evidence is in. The media rarely present the case for the accused, at least not at the beginning of a story. Once the media find out that an injustice has been done, they'll usually report it. But the initial story usually shows little sympathy for the accused. The media get their information from the accusers and often show no skepticism regarding the information or misinformation they are given by law enforcement agents. Once a case starts building, both police and journalists seek out confirmatory data. Rarely does anyone make a serious effort to exculpate the accused or consider alternative interpretations of the data. We might desire the media to be watchdogs over the police and to protect citizens from unjust accusations, but more often than not the media work with the police and present the news from the point of view of law enforcement. The practice of police and media working together can be dangerous, but it doesn't always lead to injustice. But it is cases like that of William Bennett, Edward Humphrey, and Richard Jewell—where there is strong pressure to make an arrest and to be first with the latest developments in an exciting story—that our critical thinking skills are most needed. Unfortunately, it is in just such cases that we often forsake critical thinking.

A large part of an entire community in Durham, North Carolina, abandoned critical thinking shortly after March 13, 2006. In Chapter Three I discussed the photo lineup used by police in their investigation of a charge made by a young black woman that several members of the Duke University lacrosse team had raped her. She identified three men from the lineup as those who raped her. News reporters, the district attorney, the police, some college administrators, and some student leaders did more than just state the accusation. They made it clear that they believed the woman's story and the next step was to ferret out the guilty parties from the lacrosse team. The men did not immediately deny the allegations, but that may be because they were not asked about them by any officials. The president of Duke University suspended the team while the allegations were being investigated, but he did not ask the players anything. The police stated the allegations, but did not immediately interrogate anyone. The district attorney charged three men with rape, but he did not ask them about the allegations. Each of the lacrosse team members (except the lone black player) supplied DNA for testing against a semen sample taken from the alleged victim. None was a match. District Attorney Mike Nifong did not even interview the alleged victim before filing charges of first-degree rape, kidnapping, assault by strangulation, and robbery against the three men. Why not? It appears he was taking advantage of rage in the black community that he thought might help him win an election.

Soon after news of the alleged sexual attack had been reported on local television stations and in newspapers, protests at Duke and in Durham began that lasted for several days. The protests were led by black students and black community members and the purpose of the protests was to make sure that privileged white men would not get away with raping a poor black woman. One protest took place outside the house where the alleged attack occurred. These protests took place, of course, without any knowledge about the case except that a black woman, who had been hired for the team party as a stripper along with another black "exotic dancer," had accused several white men of raping her. After Duke University President Richard Brodhead announced that he was ending the lacrosse season because it would be "inappropriate" to continue playing while the investigation went on, about 100 students approached his office and demanded a stronger response. Again, these students had no knowledge of the case except that a party had occurred at which strippers were hired to perform and that allegations of sexual assaults had been made. One of the black student leaders, sophomore Kristin High, stated: "We understand that the legal system is that you are innocent until proven guilty. But people are nervous and afraid that these people are going to get away with what they did because of a wealthy privilege, or male privilege, or a white privilege." High's sentiments seemed to reflect those of many people in the black community. After Brodhead met with students at the campus's black cultural center, graduate student Michelle Christian indicated that Duke was downplaying the alleged attack. "They need presidents, they need administrators, they need faculty, to tell them that it was wrong behavior and that they are not going to be coddled because they are athletes, because they come from privileged backgrounds, because they have money," she said.

Nifong, rather than try to defuse the explosive situation by advising restraint as the investigation went forward, instead threw fuel on the fire. "The circumstances of the rape indicate a deep racial motivation for some of the things that were done," he said—as if the accused had already been proven guilty. Duke Provost Peter Lange also threw fuel on the fire when, in response to a protest held outside of his home where protestors banged on pots and pans, he emerged and said that he believes "the students [who raped the girl] would be well-advised to come forward. They have chosen not to." David Addison of the Durham Police Department also threw fuel on the fire by claiming that "We do know that some of the players inside the house on that evening knew what transpired, and we need them to come forward." It was clear that Addison didn't want players to come forward to deny the allegations. He wanted players to testify that the allegations of rape were true and that they saw who raped the stripper. "We will be relentless in finding out who committed this crime," Addison was reported as saying.

On December 29, 2006, more than nine months after the alleged gang rape took place, the North Carolina State Bar announced that its Grievance Committee had found reasonable cause to call Nifong to its Disciplinary Hearing Commission for trial. According to the Grievance Committee, Nifong had made unethical statements to the news media about the case, such as his assertion that the sexual assault was racially motivated. He also had referred to some Duke lacrosse players as "hooligans," though on what grounds is not known. Why would a district attorney violate the rule that requires prosecutors to "refrain from making extrajudicial comments that have a substantial likelihood of heightening public condemnation of the accused"? Brian Meehan, president and director of DNA Security, a private lab in Burlington, admitted under oath that he and Nifong had agreed to withhold information favorable to the three indicted suspects. The State Bureau of Investigation lab in Raleigh also found no DNA from any of the lacrosse players in or on the accuser's body. The labs did find genetic material from other men in the accuser's underwear and body, however. Mehaan testified that he told Nifong about the DNA test results as early as April 10, 2006, a week before Nifong got indictments against Reade Seligmann and Colin Finnerty, and a month before indicting David Evans. Why would Nifong "engage in conduct involving dishonesty, fraud, deceit or misrepresentation" and "engage in conduct that is prejudicial to the administration of justice," knowing that such behavior is forbidden by the Code of Professional Responsibility? It appears he did so to gain the support of the black community, which he believed he needed in order to win an upcoming election. (He won, but his strategy backfired, as he was eventually disbarred for his malfeasance and sentenced to one day in jail for lying during a hearing.) In short, he used his position as D.A. to manipulate the media and the black community to help get him re-elected. Because of Nifong's unethical behavior, the State Conference of District Attorneys issued this statement: "It is in the best interest of justice and the effective administration of criminal justice that Mr. Nifong immediately withdraw and recuse himself from the prosecution of these cases and request the cases be assigned to another prosecutorial authority." In the meantime, three men faced charges of restraining the accuser in a bathroom and committing a first-degree sex offense against her. The rape charge was dropped when the accuser changed her story and told an investigator she wasn't sure she had been penetrated by a penis, a requirement for rape in North Carolina. As noted in Chapter Three, the men were identified by the accuser from a biased photo slide array. From information that has been reported regarding the night in question, I think it is unlikely that any jury would consider the alleged victim a credible witness. Apparently, North Carolina Attorney General Roy Cooper agreed. On April 11, 2007, he not only dropped all charges against Evans, Seligmann, and Finnerty, he also declared them innocent and victims of a "tragic rush to accuse" by an overreaching prosecutor who could be disbarred for his actions. The three men accused and maltreated by Nifong sued the city of Durham for $30 million. (Crystal Gail Mangum's charges were proved false. She has a long history of trouble with the law, the latest being an attempted murder charge for stabbing her boyfriend.)

Liberal Bias in the Media?

Manipulating the press is not restricted to local politicians or office seekers, of course. Whole nations have been manipulated by powerful politicians who control access to information that the media and the public might desire. There is little doubt that the Bush administration manipulated Congress, the United Nations, and the American people about weapons of mass destruction being transferred to al-Qaeda by Saddam Hussein for potential use against the United States. Bush's father, George Herbert Walker Bush, also manipulated Congress, the United Nations, and the American people as he led the U.S. into the "Gulf War" in 1991 after Iraq had invaded Kuwait. The public relations firm of Hill and Knowlton, headed by Craig Fuller, President G. H. W. Bush's former chief of staff, was hired by the swell-sounding "Citizens for a Free Kuwait" (CFK) to get support for U.S. intervention. The emir of Kuwait funded CFK. Hill and Knowlton ignited public opinion to go to war by coaching the daughter of Kuwait's ambassador to the U.S. to lie about herself to a Congressional committee (she said she was Nayirah, a Kuwaiti refugee) and lie about witnessing atrocities against babies by Iraqi soldiers. President Bush cited the atrocities in an address to the nation about the new Hitler in Iraq. Hill and Knowlton had unrestricted travel privileges in Saudi Arabia, while the travel of journalists was severely limited. The PR firm was the source of many amateur videos shot inside Kuwait and smuggled out to be edited and distributed on behalf of their client, the emir of Kuwait. These videotapes were widely used by TV news networks. Hill and Knowlton also coached Fatima Fahed, a close relative of a senior Kuwaiti official and the wife of Kuwait's minister of planning, for her false testimony before the United Nations Security Council about atrocities she alleged she had witnessed in Kuwait. Journalist Morgan Strong interviewed Fahed in Jedda, Saudi Arabia, before her UN testimony and she told him she had no firsthand knowledge of atrocities (Strong, p. 11). Says Strong: "It is an inescapable fact that much of what Americans saw on their news broadcasts, especially leading up to the Allied offensive against Iraqi-occupied Kuwait, was in large measure the contrivance of a public-relations firm" (Strong, p. 13).

So, what is a critical thinker to do given the prevalence of misinformation provided by interested parties who are able to hide their true origin? Be skeptical. Don't assume that what is reported is actually the case. Charges do not imply guilt. Accusations do not imply facts. Sad stories and crocodile tears don't imply that the one crying is telling the truth, the whole truth, or nothing but the truth. This is especially true when the one crying is trying to blame the media, rather than the facts that the media has reported, as the source of all their troubles. There are few better whiners about the biased media than those whose real goal is to bias the news in their favor whenever possible. The story about Vice President Cheney's interview with Armstrong Williams, which opened this chapter, is just one example of the hypocrisy of many media critics. Of all the hypocritical accusations made against the media, however, none has been more successful than the cry by conservatives of *bias by the liberal media.*

Despite the fact that the media is constantly being criticized for having a liberal bias, the evidence is overwhelming that political conservatives like Rush Limbaugh, Bill O'Reilly, Ann Coulter, and Sean Hannity perpetuate the myth of the liberal media by their incessant complaining about it and their attempt to pass off their own conservative bias as an antidote to liberal bias, an antidote that is somehow magically "fair and balanced."

The mass media are, for the most part corporately owned and exist to pursue profits for their stockholders. Like most corporations, media corporations try to present themselves in such a way as to be attractive to potential advertisers and so as not to antagonize those who have the power to legislate their activity. In 1982, fifty corporations owned almost all of the major media outlets in the United States: they owned 1,787 daily newspapers, 11,000 magazines, 9,000 radio stations, 1,000 television stations, 2,500 book publishers, and seven major movie studios. Five years later, ownership of all these media outlets had shrunk to twenty-nine corporations. By 1999, nine corporations owned the lot (McChesney 1999). The interests of the media are corporate interests. Yet, since 1971 conservatives have systematically campaigned to create the myth of the conservative media being oppressed by the liberal media (Lakoff 2004; Powell 1971). Anytime the media reports on anything that conservatives consider a hindrance to free markets and government non-interference with trade and commerce, the media can expect to be blasted for its liberal bias. The Democratic party has been painted as the "anti-business," "liberal" party, even though that party receives about ten times as much financial support from corporations as it does from organized labor (McChesney 1997). As Robert McChesney notes: "The notion that journalism can regularly produce a product that violates the fundamental interests of media owners and advertisers and do so with impunity simply has no evidence behind it. It is absurd." The evidence produced by conservatives to support the myth of liberal media bias is of two sorts. One, they note that more journalists vote Democratic than vote Republican. Two, they note that news stories more often have a "liberal" slant than a "conservative" slant. The problem with this evidence is that it identifies voting Democratic with being liberal, which is not justified. Democrats are as likely as Republicans to be conservative on issues of trade, taxation, and government spending. Also, those who accuse the media of liberal bias assume that a liberal journalist can't report anything fairly because of liberal bias, but a moment's reflection reveals this as a self-defeating criticism. If liberal journalists can't be unbiased, then neither can conservative journalists. Anyway, what does it mean to say that there is a liberal slant on climate change, airplane crashes, indictments of government officials, or mandatory drug testing?

There is bias in the media, but the bias is toward the bottom line: profits. Journalism has its own superstars, celebrities who are supposed to be able to gather unbiased information from people they interview or investigate. Their main job seems to be to attract viewers and readers. Entertainment and news are blurred together. Murder trials are televised; lawyers and judges play to the camera. Politicians and celebrity journalists hobnob together at social events. The last thing conservative critics of the media want is autonomous media. The likelihood that journalists will go en masse against the corporate interests of their bosses any time soon is very, very remote.

Bottom-line journalism has produced some biases that we should be concerned with, however. For example, news programs devote a disproportionate amount of air time to violent crimes. Consider that from 1993 to 1998, the homicide rate nationwide dropped by 20 percent. In the same period, coverage of murders on the ABC, CBS, and NBC evening news increased by 721 percent (Vincent Schiraldi, director of the Justice Policy Institute). Local news programs are similarly biased.

In the fall of 1995, Rocky Mountain Media Watch analyzed tapes of local evening news programs of 100 television stations in thirty-five states that aired the same day. Thirty percent of the news was devoted to crime. Coverage of government came in a distant second at eleven percent and environmental stories accounted for two percent of the stories covered that day. Poverty received 1.8 percent of air time. Unions and labor got 1.6 percent and civil rights got 0.9 percent. Other news of the day included a Miss Bald USA contest, a beauty contest for cows, a bourbon-tasting contest, and a story about a kangaroo who fell into a swimming pool in Australia. The motto of local TV news seems to be *if it bleeds, it leads.* Stories about crime, disaster, and war averaged forty-two percent of the news on all 100 stations. (Carroll 2004: 72)

There is definitely a bias here, but it is neither liberal nor conservative. The bias is toward cutting costs while increasing ratings.

In 1990, the *Columbia Journalism Review* published an article by John McManus, who spent 50 days inside TV newsrooms in several metropolitan areas. According to McManus, "Overall, 18 of the 32 stories analyzed—56 percent—were inaccurate or misleading." There was a pattern, too. "There is an economic logic to these distortions and inaccuracies. All but one...were likely to increase the story's appeal, help cut down the cost of reporting or oversimplify a story so it could be told in two minutes." (Carroll 2004: 72)

Questioning Our Sources

One question we might want to ask ourselves is *do we need the media anymore?* Much of the information they provide is impossible for us to check. We know the media oversimplify complex issues and appeal to our fears to attract us to their stories and thus to their advertisements. The Internet has already had a major impact on newspapers. Newsrooms are cutting back as advertising revenue shifts to online sources. Many newspapers are closing down their foreign offices and reducing what little investigative reporting they had been doing. It is possible that one day no major newspaper will have reporters in the places where major news is happening. We will have to rely solely on local providers, where fierce competition and other factors might well affect the quality of the information passed on. One day we may have to rely completely on biased organizations to provide the rest of the media with stories from around the world. Local news and reporting may be reduced to nothing but entertainment and opinion-based material. There will be no in-depth investigative reporting. Rather than being a watchdog of government, police, the military, or corporations, the media may become the protector of those agencies. The future of journalists as watchdogs may belong to the bloggers, but bloggers can be manipulated just as easily as high profile journalists.

None of us have the ability or the time to investigate everything. We are all dependent on others for information. How can we avoid being misled or manipulated? There is no formula, unfortunately, that we can follow to guarantee that we have put our trust in only those sources that deserve to be trusted. As difficult as it is to know which sources to trust, there are some general guidelines we can follow. The guidelines will vary depending on whether we are evaluating an individual, a mass media corporation, a government agent, or something as complex as the Internet. However, the rules will be basically the same whether we are evaluating a claim by a scientist, a newspaper reporter, a television or radio talk show host, a corporation CEO, a politician, or an author of a World Wide Web site.

The likelihood that a source is credible, unbiased, and accurate will depend on such things as the source's qualifications, integrity, and reputation. Does the source have the necessary qualifications for understanding and evaluating the kinds of claims he or she is making? Is there any reason to question the honesty or integrity of the source? Does the source have a reputation for accuracy? Does the source have a motive for being inaccurate or dishonest that is likely to outweigh the need to be accurate and honest?

Is the source making claims that require special knowledge or expertise? If so, does the source have that knowledge or expertise?

Sometimes the educational background of the source can help determine whether he or she is to be trusted. This is especially important if the source is making claims that only people with degrees or professional experience should be making. Sometimes knowing that a source has won prizes or has had books or articles published in his or her field of expertise that have been praised by others in the field is a sign that the source can be trusted. Winning a Nobel Prize in physics, however, doesn't qualify one to make claims in biology or climate science. Likewise, being a well-respected biologist doesn't qualify one as an expert in economics or psychology.

Does the source have a hidden agenda? Just because one is an expert, even a scientific expert, does not mean that one is necessarily above using one's position to further a racist, sexist, religious, political, or personal agenda.

Is the source paid for his or her testimony? In itself, getting paid does not taint testimony, but an expert who makes a career out of getting paid for testimony that is always favorable to the client should be looked at with a very careful eye.

Most of the time, we are at the mercy of sources who are experts in fields we know little about. We are not likely to know much about the reputation of such sources. Consider, for example, the automobile mechanic. Imagine a situation with automobile mechanics that parallels that of psychiatric testimony in the courtroom. If whenever you had car trouble it were possible to line up mechanics on one side to say the car needs new spark plugs and another group of mechanics to say the car didn't need new plugs, would you ever take your car to a mechanic? Would it make you feel more comfortable if the mechanics who always diagnosed "needs new spark plugs" were paid by The Spark Club Lobby for their opinion and those who always rejected that diagnosis were paid by The Tow Truck Society?

Often we have to rely on the testimony of others as to the reliability of a source. Knowing this, many advertisers hire actors to pretend they are satisfied customers and infomercial makers hire actors or real people to testify about the wonders of some product. What you want is to talk directly to someone who used the product or the services of the source. You want someone who is unbiased, who isn't being paid for his testimony, who isn't going to get to be on television for his testimony, and who has little to gain if you do or don't follow his recommendation.

If we are uncertain of a source's reputation, we can always seek a second opinion. Of course, experts can easily deceive us since we usually lack the knowledge and experience necessary to judge their expert opinions. However, unless deceit is common in a profession or in a particular business, the fear of acquiring a reputation for dishonesty is a major disincentive to deceive. If, on the other hand, there is little chance of getting caught and a good chance of making money by being dishonest, then the major disincentive to cheating is gone and the buyer should beware. Remember the Sears Auto Shop scandal in California in 1992. A whole area of auto repair in Sear's Shops—brake and shock absorber repair and replacement—was discovered to be fraudulent. Sears's auto shop personnel systematically lied to customers about needing brakes and shocks. They took advantage of the customer's vulnerability and the fact that there was little chance it would be found out. Unfortunately for Sears, the State of California had agents bring in vehicles with known good brakes and shocks who were told repeatedly in stores across the state that the brakes and shocks needed to be replaced. Sears took out full-page ads across the state that declared something like "mistakes were made and they will not happen again."

Many people get their information from television or websites about such matters as law, medicine, psychology, history, and investing money. While many television programs are based on facts and real cases, they are mainly produced for entertainment. One might even say that they are mainly produced to attract advertisers because many programs depend on advertising for all their revenue. Just because a network is called "The History Channel," however, does not mean that it cares one way or the other about historical accuracy. Internet health sites that mostly promote the sale of supplements or natural products, or try to scare you about all the toxins in your body and all around you, may look slick and provide what appear to be reliable and informative medical articles. Usually such websites are one-sided and have a specific agenda such as selling supplements or opposing vaccinations.

Just because you saw or heard it on "Law and Order," "Dr. Phil," or "The Dr. Oz Show" does not mean that you've been given sound legal or medical advice. Even PBS programs cannot be given a blank check for credibility. Some PBS programs are infomercials produced by their stars. They are not produced by PBS or any of its affiliates. For example, "Magnificent Mind at Any Age with Dr. Daniel Amen" and "Change Your Brain, Change Your Life," were produced by Dr. Amen himself. In the programs, he makes unsubstantiated claims about natural products and supplements that he just happens to sell from his website. I don't care if a private station puts on an infomercial, but I care if my tax dollars are used to make an infomercial available to a large public television audience. Why should the taxpayer subsidize these programs? And why do we let PBS get away with the ruse of "underwritten by" or "with support from" when these are just paid advertisements by another name?

At least we know who is paying for those swell commercials that aren't commercials on PBS when we're told that a program about saving the environment for future generations was made possible by contributions from Exxon or Mobil Oil or David H. Koch. Many times, though, important news comes from *anonymous* sources. It wouldn't be fair or wise under most circumstances to allow people to testify against others in court anonymously. Often, however, the only way vital information will ever be released to the public is under the condition that the source remains anonymous. There are times when naming a source would be dangerous or potentially harmful to the source, but nobody should be given a blank check to make serious allegations about others just by claiming we should trust them because they got their information from an unnamed reliable source. Unfortunately, even when we know the sources and they seem to be trustworthy, the truth often turns out to be quite different from what it at first appears to be. This is true even in the sciences, a subject we will take up in the next chapter.

SOURCES FOR CHAPTER SIX 273

VII
Seductive Stories and Varieties of Scientific Experience

"If your experiment needs statistics, you ought to have done a better experiment."—Ernest Rutherford

"Statistical significance can be totally meaningless and it usually is."—James Alcock

---

Never let truth get in the way of a good story. That seems to be the motto of not only many journalists but many scientists as well. A good story trumps a dozen scientific studies. Stories provide anecdotal evidence based on personal experience and are sexier than studies. But, as we saw in Chapter Three, experience is hardly an infallible guide to reality. Anecdotes are powerful persuaders, but they are not compelling evidence to a scientist.

Scientific tests are supposed to avoid the pitfalls of relying on anecdotes, but scientific studies are often unreliable or misleading. Scientific studies can mislead us with statistics, graphs, or questionable analogies. They can be biased or poorly designed. Scientists can draw strong conclusions from small samples. Data can be interpreted incorrectly. Scientists can use inadequate controls or no controls at all. Studies can have high attrition rates that can skew the data. And, unfortunately, there is the occasional case of fraud.

We often draw drastic conclusions about how humans might live better or longer lives based on a single study involving rodents in a lab or cells in a test tube. Science may be evidence-based but that does not exempt it from the effects of the numerous cognitive, perceptual, and affective illusions and biases we have considered in earlier chapters. In this chapter, we'll look at a few more biases: experimenter bias, expectancy bias, the clustering illusion, the file-drawer effect, the placebo effect, and publication bias.

Fear Sells

On the morning of September 14, 1989, Joseph Wesbecker, armed with an AK-47 and a handgun, entered Standard Gravure, a printing company in Louisville, Kentucky, where he had worked for seventeen years. He shot twenty of his co-workers, killing eight of them. He then committed suicide. In the spring, Wesbecker had been put on disability leave while being treated for a mental illness. For about a month before his deadly outburst, Wesbecker had been prescribed Prozac, a drug approved by the FDA in 1987. The families of those killed by Wesbecker, along with those who survived the deadly rampage, sued the manufacturer of Prozac, Eli Lilly and Company. Perhaps it was just coincidence or perhaps it was at the behest of attorneys for the plaintiffs, but Wesbecker's sons went on the television talk show *Larry King Live* and one of the surviving victims went on the *Phil Donahue Show* to assert their conviction that Prozac caused Joseph Wesbecker to become violent and murderous. Donahue's program was called "Prozac: the medication that makes you kill." At the time of these allegations regarding the lethal effects of Prozac, there was no compelling scientific evidence that Prozac, or any other serotonin uptake inhibitor, is a significant causal factor in violent behavior, much less a trigger for deadly rampages. There still isn't any compelling scientific evidence linking Prozac to murderous or suicidal inclinations, though there is strong evidence that such drugs can delay sexual climax and could be just what the doctor ordered for those suffering from premature ejaculation. There haven't been enough large-scale clinical trials using double-blind, randomized controls to justify claiming there is strong evidence of violent behavior being caused by Prozac. The fact that Wesbecker's victims lost their lawsuit does not prove that Prozac is safe or effective as an anti-depressant, however. A lawsuit, regardless of the outcome, is just one more anecdote as far as scientific evidence is concerned. The plural of 'anecdote' is not 'data.'

Because of fears aroused regarding the safety of antidepressants like Prozac, there was a sharp decrease in the prescribing of such drugs after the Wesbecker media frenzy. Robert Gibbons, a professor of biostatistics and psychiatry at the University of Illinois at Chicago, reported in 2007 that the decrease in use of antidepressants was *not* followed by a decrease in suicides. In fact, just the opposite happened: the suicide rate went up as antidepressant use went down. In the Netherlands, for example, there was a 22 percent *decrease* in antidepressant use between 2003 and 2005, yet the suicide rate for those under nineteen years of age *increased* by 49 percent. Even though several studies have shown that where antidepressant use is high the suicide rate is low, the media and talk show hosts focus on cases of suicide or attempted suicide where antidepressants have been used. Clearly, not everyone with suicidal inclinations will be calmed by antidepressants, but to blame the antidepressants for Wesbecker's suicidal rampage was unwarranted.

Talk shows are particularly effective in promoting beliefs based on anecdotes not supported by scientific evidence. When she was in her prime as a talk show hostess, Oprah Winfrey was the queen of the good story and the anecdote that substitutes for serious analysis. In less than an hour she could turn a minor tale of something like "road rage" into a candidate for admission into the Diagnostic and Statistical Manual of Mental Disorders. I know Winfrey has done some wonderful things in her life, but that shouldn't blind us to the negative side of her powerful personality. She can help women win baseless lawsuits just by parading a few sick ladies across the stage while noting that they all had breast implants. Of course, Oprah is just one of many in the mass media who play on fear and use questionable authorities and statistics to back up allegations of plagues and epidemics of everything from road rage to Internet addiction to vaccines causing autism.

Many people are convinced that silicone breast implants have been shown to cause all kinds of physical diseases and disorders because, after all, many women got sick after their implants and Dow Chemical filed for bankruptcy in 1995 as a result of the many lawsuits against it by alleged implant victims. The lawyers extracted several billions of dollars in a settlement against the implant manufacturers without providing compelling scientific evidence that the implants were harming women. Two seemingly reputable scientists testified as experts for the plaintiffs, but no scientific studies were produced because there hadn't been any such studies yet.

Many were convinced of the causal connection between implants and disease when they witnessed talk shows during the 1990s with woman after woman telling her story of getting an implant and then later suffering from cancer, lupus, migraines, and other disorders. Ralph Nader's Public Citizen Health Research Group sent out a warning that silicone breast implants cause cancer. The FDA banned silicone breast implants in 1992. The media, consumer advocates, talk show hosts, government agencies, and lawyers were hammering home the same message, so it is not surprising that many people still think that silicone breast implants cause all kinds of health problems. In 1999, the Institute of Medicine (part of the National Academy of Sciences) released a 400-page report written by an independent committee of thirteen scientists that showed silicone breast implants may be responsible for some localized problems like hardening or scarring of breast tissue, but they do not cause any major diseases. In 2006, the FDA lifted its ban on implants. A four-year follow-up study by the FDA released in June 2011 reiterated the fact that silicone breast implants do not cause cancer, fibromyalgia, rheumatoid arthritis, or a range of other ills. Of course, there may be cases where the surgery was botched and the patient suffered as a consequence of implant surgery. But the harm in those cases is due to the surgeon, not the silicone implant. Despite these corrections, there is still widespread belief that silicone implants cause grave illnesses. In fact, a number of psychological studies have found that "denials and clarifications, for all their intuitive appeal, can paradoxically contribute to the resiliency of popular myths" (Vedantum 2007: 5). This psychological phenomenon, known as the *continuing influence effect* differs a bit from the backfire effect, discussed in Chapter Two. With the continued influence effect, one learns "facts" about an event that later turn out to be false or unfounded, but the discredited information continues to influence reasoning and understanding even after one has been corrected. The backfire and continued influence effects should be disheartening to those who think that the first step in arguing with those who base their beliefs on misinformation should be to get their opponents to see what the facts are. Correcting errors may be pointless when dealing with some people. Critical thinkers, one would hope, would want errors corrected. At the very least, getting the facts right might prevent some faulty inferences and prevent one from behaving in ways that could prove harmful. Getting the facts right, however, often doesn't have the emotional or cash value that telling a good story has. Also, correcting errors is a waste of time if the one you are correcting attributes his own beliefs to principled, unprejudiced inquiry, while attributing the beliefs of those who disagree with him to bias and ulterior motives. There are few cases more exemplary of this problem of attributing evil motives to those who disagree with you than the case of those who are absolutely sure that cell phones cause brain cancer.

In 1993, David and Susan Reynard began making the talk show circuit with their claim that her brain cancer had been caused by her cell phone, an instrument she had been using for *only three months* before the cancer was diagnosed. Would Reynard have been taken seriously if he had claimed that his wife's lung cancer was caused by smoking, a habit she had taken up three months earlier? I don't think so. Anyway, David Reynard went on *Larry King Live* and told the story that the tumor was located next to where she held the phone. He lost his lawsuit. Seven years later Reynard returned to a *Larry King Live* show that featured Dr. Chris Newman, who had been diagnosed with brain cancer and was suing Motorola and Verizon for $100 million in compensatory damages and $700 million in punitive damages. Unlike the first show, however, the show on August 9, 2000, was much more balanced and featured several panel members who asserted that there was no compelling scientific evidence that cell phone use causes brain cancer. When asked how he knew his cell phone had caused his cancer, Newman replied that he *believed* it was so and then deferred to his attorney, Joanne Suder, who was in the studio with him. She said that "there really is no question about it" because Newman's "own physicians made the correlation between his longtime cell phone use and his cancer." What was their evidence? Newman had "nine years of a vast amount of cell phone use and his terminal tumor is located in the exact anatomical location where the radiation from the cell phone emitted into his skull." No appeal to compelling scientific evidence was provided. Why? Such evidence doesn't exist. That does not mean that it will never exist, but it does mean that up to now (November 2011) the data do not support the claim that radiation from cell phones poses a health hazard to the humans who use them. There have been a number of large studies on EMF exposure and the preponderance of the evidence is that there is no great danger to humans from using cell phones, Wi-Fi, or living near power lines. (The latest study on cell phones—published in October 2011—was done in Denmark and involved over 350,000 cell phone users over ten years; there was no difference in cancer rates among users and non-users.) For more details on the cell phone and related scares see Appendix A.

In our quest to discover the causes of a disease or an action, there are sound reasons for preferring the data from randomized, double-blind, controlled experiments to the data provided by anecdotes. Even well-educated, highly trained experts are subject to many perceptual, affective, and cognitive biases that lead us into error when evaluating personal experiences. The informal procedures most of us use to decide whether events are causally related are vastly inferior to formal rules such as Mill's methods (some of which we will describe below). The formal rules aren't infallible and can be misapplied, but they are orders of magnitude more reliable than naïve sense perception, unaided memory, or intuition. Formal methods of causal analysis are necessary even if we do not have an emotional, doctrinal, or monetary stake in the acceptance of a particular causal claim. While it's true that many of our beliefs are driven by our biases and are generated for their comfort-value rather than their truth-value, formal methods of causal analysis are necessary even if the causal claim is not particularly comforting or attractive. All things being equal, the more impersonal and detached we are in evaluating potential causal events, the less likely error becomes.

Thus, when personal experience conflicts with the data of well-designed and executed control-group studies, it behooves us to examine our personal experience for bias. We should be especially wary of any products promising beneficial effects that are marketed before control studies have been done. For example, magnetic devices on the market today that claim to be beneficial for relieving pain have not been rigorously tested. It is not unusual to find a single study here or there that supports just about any causal claim you can think of, but single studies rarely, if ever, provide compelling proof. When there is no compelling supportive evidence from control studies, we should subject a causal claim to a detailed critical evaluation before concluding that there is a causal connection between two events. For example, the evidence from control studies indicates that the claims of applied kinesiology (AK) are bogus. (AK uses manual muscle resistance testing to determine a host of unrelated things like illness, truth, the goodness of a sugar, or the power of a rubber band bracelet.) This fact has not diminished the attractiveness of AK for many people. Remember the chiropractors mentioned in Chapter Two who rejected randomized, double-blind, control-group experiments on the ground that they don't work, i.e., the results conflict with their beliefs about AK. Of course, the scientific studies could be wrong, and it would be wise to evaluate them carefully before rejecting your conclusions drawn from personal experience. But the response of the AK chiropractors was irrational. Instead of rejecting formal, scientific testing, a rational person would inspect his conflicting belief and make every effort to discover the source of his error or attempt to discover flaws in the scientific studies.

Anecdotal Evidence and Scientific Studies

The benefits of formal analysis and the pitfalls of anecdotal evidence are illustrated in the following e-mail I received from "Mike":

Twenty-five years ago (at age 30) I began suffering from serious food allergies. I began reacting to common foods that I had eaten for most of my life with increasing severity including a couple of potentially life threatening anaphylactic shocks that resulted in my being rushed to nearby hospital emergency wards and saved with injections of epinephrine.

I exhausted the best traditional medical resources available in Boston without results. Out of desperation, I visited a practitioner of applied kinesiology on a recommendation from a local nutritionist. It was the strangest thing. This very professional female in a white lab coat had me lie down on a medical-style table in what looked like a traditional doctor's office. She placed dozens of small amounts of food substances enclosed within what looked like small plastic pill boxes on my stomach while I lay face up. She asked me to extend my right arm and resist her efforts to press my arm down toward the ground. In most instances, I easily resisted her pressure. In some instances my arm collapsed, which was astounding to me at the time both physically and mentally. She meticulously created a list that included several food substances to eliminate from my diet. I did so.

From that day forward I stopped having allergic food reactions. After a few months I visited her again. She tested me for the items I showed sensitivity to during our first visit. I tested as cured. Cautiously, I reintroduced those food items back into my diet without any repercussions. I have been fine ever since. I don't know how or why this worked. I do know that very sophisticated lab work conducted in a highly regarded hospital in Boston by a renowned physician was unable to determine what I was allergic to and how to pursue relief.

Mike says he doesn't know how or why AK worked. Mike seems to think that AK worked better than the best science available at the time since the AK therapist found what caused his allergies and the doctors using scientific medicine did not. However, from this anecdotal evidence alone we can't justifiably say that AK worked at all. It is an *assumption* that Mike suffered from food allergies identified by the AK therapist.

It is natural that a person suffering from some problem would infer that since the problem went away after the treatment that the treatment was effective. However, the only evidence we have that the treatment worked (i.e., was effective) is that the problem went away *after* the visit to the AK therapist and *after* Mike stopped ingesting the foods she told him to avoid. But lots of things happen *after* other things without causing them. I raise my hand and right afterward a lightning bolt lights up the sky. Did I cause the lightning? No. We've all met people who are convinced that Echinacea or some other herb prevents colds because every time they feel a cold coming on they take the herb and "it always works." The stories may be true, but there are many people who *don't* take herbs every time they feel a cold coming on and they haven't had a cold lately, either. Is that evidence that *not* taking Echinacea or some other herb prevents colds? For those who like Latin, reasoning that x caused y because y came after x is called the *post hoc fallacy*. Post hoc is shorthand for *post hoc ergo propter hoc—after this therefore because of this*. For those who prefer the data of scientific studies to the data of personal experience, the National Institutes of Health has funded two studies on Echinacea. Neither found supportive evidence that the herb—either as *Echinacea purpurea* or as an unrefined mixture of *Echinacea angustifolia* root and *Echinacea purpurea*—prevents or lessens the severity of colds in adults or children.

The fact that conventional medical tests did not indicate that Mike's symptoms were due to an allergy is not evidence in favor of the view that AK is superior to conventional scientific methods. Had Mike not visited the AK therapist, his symptoms may have ceased without any intervention at all.

Mike's anecdote illustrates why scientists prefer controlled studies to the details of personal experience, even though those details are important for medical diagnosis. The first problem with the anecdote is that we have to rely on Mike's memory for details. The events happened twenty-five years ago. A lot of things can conspire to corrupt even a recent memory. The accuracy of the details may not be reliable. Memory can be selective and important details may not be recalled. Things that Mike experienced in the past twenty-five years might get mixed up with the old memory. Time sequences can get confused. A properly designed and executed control study can eliminate these problems.

Then there is the problem of interpreting the data. Mike seems convinced that he suffered from food allergies. Hypothetically, it would be relatively easy to test AK's effectiveness at diagnosing food allergies. First, we draw up a list of allergens and non-allergens for the patient based on whatever standard methods are used to measure such things. This assumes, of course, that there are such methods. We label the allergens and non-allergens with different letters of the alphabet. Let's say we have a list of twenty items, ten allergens and ten non-allergens. We use a randomization procedure to select, say, ten items and put them in ten labeled pill boxes. The boxes must be opaque and shut tight, so that neither the client not the AK therapist can see what's in them. All samples should weigh about the same to prevent the weight of the item revealing to the client or the AK therapist what it is. There should be no other obvious telltale sign, like odor, to tip off the therapist. One researcher should put each item on the list in a pill box and record which substance is in which box. A different researcher should give the boxes to the AK therapist who can then use the arm pressure test with the pill box on the subject's stomach. Such a test is called a double-blind, randomized test. The items to be tested have been chosen in an unbiased way and neither the subject nor the AK therapist knows what is in the box being tested. When the results are all in, the test can be unblinded, i.e., it can be revealed to subject and therapist what was in each box in each test. We then compare the AK therapist's decision (allergen or non-allergen) with the known facts.

Why do it this way? Because the client or AK practitioner might react differently if they know what is in the box being tested. If the subject knows the item is an allergen, he might unconsciously weaken his resistance. If the therapist knows the item is not an allergen, she might unconsciously decrease the pressure she applies.

Since there are only two possibilities for each item (allergen or not), there is a 50% chance of being right each time. If our therapist gets ten out of ten correct, we would be justified in saying that such a result was not likely due to chance. If we do hundreds of tests with dozens of subjects and several trained AK experts and overall the results are significantly greater than expected by chance, then we would be justified in concluding that AK is a reliable method for detecting food allergies.

Unfortunately, with Mike's anecdote we have no way of knowing whether the AK practitioner's pill box routine reliably identifies food allergies. The list she gave Mike could have been determined by her pill box routine, but the amount of pressure she exerted for each test might have been unconsciously determined by her beliefs about the allergenic potential of what she knew was in the pill box for the test. Even more unfortunate is the fact that there is no compelling scientific evidence to support the claim that this AK practitioner's method of identifying food allergens is reliable.

Some readers might be impressed by the fact that Mike's symptoms went away after his visit to the AK therapist. Doesn't that prove she cured him? Not really. We can't be sure that the absence of further symptoms was due to the absence of the foods removed from Mike's diet. For all we know, some unknown factor (an X-factor) caused the anaphylactic shocks, some event or substance not controlled for and completely independent of any of the foods Mike believes caused his symptoms. The best medical people in Boston couldn't identify the cause of Mike's symptoms, he tells us. The belief that food caused his medical problems seems based on little more than the claims of the AK practitioner and the fact that his symptoms did not recur after visiting her and he stopped eating the foods named on the list she gave him.

Some might say "what harm is there in AK? Mike got better and that's all that matters." While it certainly matters that Mike got better, that is not *all* that matters. We don't have any anecdotes from the unsatisfied customers, the ones who either died or recovered on their own, or were suffering from a treatable disorder and who received proper care from a proper doctor and recovered because of that care. One of the problems with anecdotes is that they tend to be provided by the satisfied customers, not the unsatisfied or dead ones.

It is puzzling, however, that even though he received no therapy, he was told by the AK therapist several months after his first visit with her that he was cured of his allergies and could now return to eating the foods that nearly killed him. He says he returned to eating these foods and hasn't had any more problems. This evidence is consistent with the claim that an X-factor, not the foods he avoided for a few months, caused his symptoms. This anecdote, like most anecdotes, can be interpreted in several plausible ways. A properly controlled series of studies would produce less controversial results. Perhaps Mike can now eat the killer foods because he never had an allergic reaction to them in the first place. Perhaps his symptoms weren't caused by an allergic reaction to anything. Something else that we know nothing about may have caused his symptoms. Whatever it was, it was temporary. Maybe stress or an infection caused the symptoms. Maybe the placebo effect can best explain why he got relief after visiting the AK therapist. (The placebo effect is the measurable, observable, or felt improvement in health not attributable to an active treatment.) I am no expert on food allergies but from what I do know it seems unlikely a person could go thirty years without a reaction to certain foods, and then suddenly be nearly killed by substances that have a long history of benevolence, and then a few months later find that the killers have returned to their former gentle selves.

This anecdote illustrates the power of experience to deceive us into thinking we have identified a causal agent when, in fact, we have very little evidence in support of our belief. Even if AK has some sort of placebo effect, that is not a good enough reason to recommend it. It is possible, I suppose, that the AK therapist was so charming and sweet that she completely disarmed Mike and relieved him of all stress, thereby relaxing him to such an extent that his own body was able to produce natural substances that brought him relief or shut down the production of the harmful chemicals that were causing his symptoms. However, even if AK has a placebo effect, we have no way of knowing this without doing systematic formal testing of AK. We should remember, too, that most of the things that ail us will not kill us. Most diseases are self-limiting. The human immune system provides most of us with a recovery mechanism for most of our ailments. As the saying goes, *what doesn't kill me makes me stronger.*

It might seem that the evidence in Mike's anecdote that AK is an effective diagnostic and therapeutic procedure is on par with the evidence that the prescription a physician gives you, say, for a diagnosed bladder infection is effective. It might appear that the only evidence you have for the effectiveness of the medicine is that the infection went away after you took the medicine. However, the physician chose the drug to prescribe based on compelling evidence from scientific studies. The AK therapist has no comparable compelling evidence from scientific studies to back up her method of diagnosis or prescribed treatment. What she probably has is some experience using AK where the method seems to work. She has probably seen demonstrations of AK, has received some positive feedback from satisfied customers, and so "knows from experience" that it works. The evidence from satisfied customers is anecdotal and consists of nothing more than that some customers claim that they felt better after the treatment. Again, these customers have no way of knowing that it was the AK that brought about the desired outcome. All they know is that one thing happened after another. The demonstrations of AK that the practitioner has probably witnessed or produced herself were most likely not controlled demonstrations. The method of testing AK used by practitioners would probably involve knowing what substance is being tested and then behaving in an expected manner (*expectation bias*). If the substance is believed to be bad for you, the practitioner presses hard or produces little resistance (depending on what role they are playing). If the substance is believed to be good for you, then the practitioner doesn't press hard or the subject doesn't offer much resistance. This may all take place at the unconscious level, however. So, we need not accuse the AK folks of fraud. Ideomotor action—unconscious muscle movement—and a strong desire to succeed in providing support for their belief system are sufficient to explain how this kind of self-deception works. These factors also illustrate the importance of doing randomized, double-blind, controlled studies rather than relying on personal experience and anecdotes to establish causal connections.

Nowhere is the need for doing properly controlled scientific studies more apparent than in the fields of parapsychology and CAM—complementary and alternative medicine—the marketing term used by those favoring treatments that aren't science-based or that have not been shown to be effective. Those fields have been dominated by anecdotes for centuries. In recent years, however, there have been many scientists investigating paranormal and non-conventional medical claims who recognize that anecdotes will never provide enough evidence to satisfy skeptics of the reality of extrasensory perception, psychokinesis, survival of consciousness, or the effectiveness of shark cartilage or energized water in curing cancer. Scientists know that collecting seemingly inexplicable anecdotes does not suffice for scientific evidence, even if the anecdotes number in the millions and even if the storytellers are Nobel Prize-winning anointed saints. Such a process puts too high a premium on our ignorance and our laziness. If it's the truth we're after, it will not serve us well to turn to paranormal, supernatural, or magical stories every time we can't think of a naturalistic explanation for an event. Evolutionary biologist and science writer Richard Dawkins calls this "lazy thinking." In science, it just won't do.

My favorite example of lazy thinking comes from noted parapsychologist Charles Tart who, ironically, provides an example of lazy thinking while arguing against it. At a conference in Casper, Wyoming, in the 1980s or early 1990s Tart repeated a story told to him by a woman who worked in his lab. On July 17, 1944, she had awakened in the middle of the night in her room in Berkeley, California, "overwhelmed by a feeling of absolute horror. She knew that something absolutely horrible was happening that she desperately wished she could stop, and she didn't have the slightest idea in the world what it was." About a minute after she had been awakened, the window and room shook. She found out the next morning via a radio broadcast that at 10:18 PM in Port Chicago, California (thirty-five miles northeast of San Francisco) a ship with more than four thousand tons of ammunition exploded, followed by sixteen rail cars with more than four hundred tons of explosives. Three hundred twenty people were killed, most of them African-Americans in the U.S. Navy. "She felt in retrospect that somehow some part of her mind had reacted to the horror of all those people dying simultaneously," said Tart. He told the story to give the audience an example of the kind of thing that leads many people to believe in the paranormal. According to him, this event is not explicable by the known laws of physics and is an example of precognition: knowing something before it happens and knowing it by paranormal means. James Randi, who heard Tart's explanation, offered one of his own. Shock waves travel at different speeds through different mediums (earth, air, water, and so on). Given the distance of Berkeley from Port Chicago, there should have been a difference of about eight seconds between the time the shock wave through the ground arrived and when the shock wave through the air arrived. The shaking ground woke up the young lady and eight seconds later the window shook. Tart's reaction was to tell Randi: "That may be the explanation that you prefer." To Tart, which explanation one accepts depends on what one already believes. He is probably right about that, but that doesn't make all explanations equally plausible. In any case, he recognized that anecdotes are problematic and will not satisfy the skeptic. "Let's take this into the laboratory," Tart said, "where we can know exactly what conditions were. We don't have to hear a story told years later and hope that it was accurate. We can keep accurate records of exactly what went on at the time and know whether we have something unusual happening" (James Randi, personal correspondence, including a transcription of a tape of Tart's talk).

Anecdotes are problematic because the stories might be contaminated. Most stories get distorted in the telling and the retelling. Events get exaggerated. Time sequences get confused. Details get muddled. Memories are selective and often filled in after the fact. People sometimes misinterpret their experiences. Some people make up stories. Some stories are delusions. Sometimes improbable events are inappropriately deemed psychic. And some experiences are deemed extraordinarily improbable when in fact they're not.

Science and the Media

Scientists aren't perfect, however. Some do randomized, double-blind, controlled studies that aren't worth the test tubes they're done in. Some scientists seem more motivated by money and fame than by truth. Some do good studies, but they draw conclusions from them that aren't justified. Some science covers areas where it is simply not possible or would be unethical to do randomized, double-blind, controlled studies. It seems fair to ask, "Just how good is the evidence in such areas as climate science, archaeology, astronomy, or the healing effects of intercessory prayer?" A more important question for most of us, though, might be: "How good is science reporting in the media?" Most of us don't know enough about science, nor do we have the time to read scientific journals. We rely on the media for our science. One obvious problem with this practice is that members of the media are usually no better trained in science than we are and are selective in what they report on and in how they report it. You'll rarely see or read a story in the media about the latest goings on in a parapsychology lab unless there is something sexy to the story like inducing out-of-body experiences in people. The more complicated the science, the less likely a journalist is going to get it right, if indeed the story is even reported on. The more complicated the implications of the science, as in climate studies, the more controversy. Controversy sells, even pseudo-controversy.

The media often give us the false impression that scientists are equally divided on an issue when they are not. In an attempt at journalistic balance or in an effort to make a story more compelling, journalists can give the public the impression that a single individual or a small band of contrarians represent many scientists who disagree with many other scientists. Cultural anthropologist Christopher Toumey calls this media practice "the pseudosymmetry of scientific authority." It can give the impression that a scientific consensus has not been reached and that an issue is considered controversial within the scientific community when, in fact, it is not.

Examples of the pseudosymmetry of scientific authority can be found in the way the American media have covered acupuncture, climate change, cold fusion, corn as the source of tomorrow's fuel, evolution and intelligent design, a human mission to Mars, free energy issues (perpetual motion machines), and fluoridation of municipal water supplies. The anti-fluoridationists continue to seek any scientific study that remotely indicates there might be some harm to somebody from fluoridation. One tactic they use is to speculate that there are some people who are "especially sensitive" to fluoride. Unfortunately, they have no way of identifying who these hypothesized sensitives might be. It must be granted that it is impossible to prove that any given substance could never harm anyone. It is always possible that there are "especially sensitive" individuals and that there is no way to identify them, e.g., by genetic markers. But even granting this, it doesn't follow that a substance should be banned just so we can be on the safe side. That kind of caution would justify banning everything! Nor does the speculation about especially sensitive people justify the claim of any individual with a disorder that it was caused by a substance to which he is "especially sensitive." Otherwise, anybody could claim that anything caused any ailment they have. But I digress.

The pseudosymmetry of scientific authority significantly affects public opinion in some cases but not in others. Pseudosymmetry has significantly affected the issue of vaccination to the point where many intelligent, educated people are not having their children vaccinated against diseases like measles. On the other hand, despite the "teach the controversy" campaign of the intelligent design folks, these anti-evolutionists have lost all major battles in court. Creationists have won a few skirmishes in the attempt to get their religious views of creation taught in the biology classroom, but they have had very little success in convincing the media or the general public that evolution represents moral anarchy and all things evil. Anti-fluoridationists, on the other hand, continue to wage many successful campaigns in public referendums, even though there is little science or fact to support their fears. The anti-fluoridationists have been helped in their cause by the widespread mistrust of government, a mistrust that fuels many conspiracy theories, including the 9/11 conspiracy movement and the AIDS denial movement. Government mistrust is evident in whatever following these movements have, but the mainstream media ignore the AIDS deniers and those who claim the Bush administration was behind the attacks on 9/11. Also, even though many media critics believe that climate change contrarians have been given equal time to consensus scientists who believe the evidence proves beyond a reasonable doubt that our planet's temperature is rising and that humans are contributing to the rise with increasing carbon dioxide emissions, 85% of us believe global warming is happening and threatens future generations (*TIME* magazine/ABC News/Stanford University poll, March 2006). And even though more than half of all American adults reject evolution, religious training and scientific illiteracy—not media influence—are likely the main factors in such massive disbelief of this basic biological fact.

Journalists are rarely trained as scientists, so it is not surprising that they covered the cold fusion fiasco of Pons and Fleischmann at the University of Utah in such a way as to give the scientists all due respect even though it turned out that these chemists (and a few others who agreed that maybe the Utah pair were on to something really big) didn't understand enough physics to know what they were doing. On the other hand, journalists should have been skeptical of the way the university handled the matter and lobbied for funds in Washington before a single scientific paper had been presented. Journalists may not understand the science but they should understand greed and deception and how they might affect competition for potentially lucrative patents. In any case, today only a few outcasts give any serious attention to cold fusion. This fact did not deter Scott Pelley of CBS's *60 Minutes* from doing a segment on 19 April 2009 that amounted to a promotional piece for the outcasts. The most pathetic segment of Pelley's panegyric to cold fusion scientists was his visit with Martin Fleishmann where he tried to cheer up the apparently defeated and destroyed old man by convincing him that he was right after all and is being vindicated. Pelley did interview one skeptic, Richard Garwin, a man who helped design the most successful fusion experiment of all time: the hydrogen bomb. He told Pelley the reason cold fusion scientists don't get measurable heat all the time is that they're deluding themselves. Their results are due to their equipment, not *cold fusion* (or to "low energy nuclear reaction," as defenders call it these days). In short, the cold fusion work isn't *more than junk science* (the title of the *60 Minutes* segment.)

Science and Politics

Unfortunately, sometimes the outcasts and oddballs who cling to ideas that the vast majority of the scientific community thinks are foolish happen to be powerful politicians whose every word, no matter how uninformed, is considered newsworthy.

The National Center for Complementary and Alternative Medicine (NCCAM), originally called the Office of Alternative Medicine (OAM), came into being because a powerful politician believed in something the scientific community considers hogwash. In 1991 Iowa Sen. Tom Harkin got the bug for so called complementary and alternative medicine (CAM) when he came to believe that bee pollen cures hay fever. Harkin was then and still is (in 2011) a powerful figure on the appropriations subcommittee in charge of the National Institutes of Health (of which the NCCAM is a part). He and a few Senate buddies wanted to fund research that would prove the effectiveness of bee pollen and other questionable therapies. There is no evidence in the scientific literature that bee pollen cures allergies or has any beneficial effect. Worse, bee pollen can cause life-threatening allergic reactions in some people. Nevertheless, Harkin and the promoters of CAM wanted the NIH to do the science to prove the benefits of specific CAM treatments. When the OAM couldn't come up with any good science for even one CAM therapy during its first eighteen months of operation, Harkin attacked the OAM's director Dr. Joseph Jacobs in a public hearing (June 1993) and then "handpicked four alternative-medicine advocates" and had them appointed to the OAM's advisory board (Satel and Taranto: 1996).

Jacobs called the CAM advocates "Harkinites" and they soon attacked Jacobs for trying to set up proper scientific research centers. Such fact-based research would be "hostile" to CAM, they said. The Harkinites won out. The OAM set up research centers at the University of Minnesota's Center for Addiction and Alternative Medicine Research and at Bastyr University, a naturopathic college outside of Seattle. The Harkinites were so resistant to sound science that Jacobs resigned in September 1994. "It's pathetic," he said. "They were so naïve about science. I wouldn't trust anything coming out of the OAM as long as the Harkinites are micromanaging it." A few years later it was announced that Congress had approved $50,000,000 to establish a new National Center for Complementary and Alternative Medicine. In 1993, the maker of the bee pollen capsules that Harkin considers a cure for hay fever—Royden Brown of the CC Pollen Co.—paid the Federal Trade Commission $200,000 in a consent decree for making false claims about his product's curative powers.

The Harkin experience is a good example of why we should not trust instinct or intuition on complex causal matters. Harkin was so sure of his position that he convinced himself that the scientists who couldn't prove him right must have been incompetent. A critical thinker would have reassessed his position and gone with the science rather than with what he "knew from experience to be true."

When it comes to journalists and politicians mediating our science, there are other problems for us to consider in addition to the problem that plagues many laypeople: putting our interpretations of personal experience ahead of the evidence from scientific studies. Journalists and politicians may not know their science or think logically but they know fear and they know that fear sells and gets votes. The media have a soft spot for science that has a scary aspect to it, as a very clever British physician figured out to his advantage and to the great disadvantage of many children.

Abusing Science and Manipulating the Media

Dr. Andrew Wakefield sounded the alarm a few years ago about a possible connection between the MMR vaccine and autism in children. Most scientists have dismissed Wakefield's work as inadequate and dangerous, even though it first appeared in the respected medical journal *Lancet*. It was inadequate because his conclusions were based on a sample of only twelve children, far too few from which to justify drawing any meaningful generalizations. A measles epidemic in Ireland (over 1,500 reported cases in 2000) has been blamed on Wakefield. Three children died. Fears of an epidemic in Scotland (where Wakefield once practiced medicine) and England were also fueled by Wakefield's claims. Cases of measles in England rose to a 20-year high following the collapse in MMR immunization rates.

We now know that Wakefield was paid more than £400,000 by lawyers trying to prove that the vaccine was unsafe. The payments were part of £3.4 million distributed from a legal-aid fund to doctors and scientists who had been recruited to support a now-failed lawsuit against vaccine manufacturers. Journalists rarely investigate the funding sources of scientific research and studies that may be biased because of funding interests are presented to the general public without any indication that the data or inferences from the data may be tainted. Usually, the expected bias would be in the form of, say, the sugar industry funding a study on the nutritious benefits of their product. It came as a surprise to many people that Wakefield was getting money from people who sue other people and companies. With the support of scientific studies that show vaccines are unsafe, lawyers would be assisted in their efforts to convince judges or juries that their client should be compensated. A similar kind of ruse occurred in 1991 when the CBS Evening News showed a video news release from the Institute for Injury Reduction that claimed automobile safety belts are unsafe. The Institute for Injury Reduction is a lobby group largely supported by lawyers whose clients often sue auto companies for crash-related injuries (Carroll 2005: p. 67).

Wakefield's misconduct was uncovered by investigative journalist Brian Deer. Deer also discovered that in 1997 Wakefield had applied for a patent for a measles vaccine on behalf of the Royal Free hospital medical school and the Neuroimmuno Therapeutics Research Foundation, a private company of unconventional immunologist Professor H. Hugh Fudenberg of Spartanburg, South Carolina. (Fudenberg claimed in a 2004 interview with Deer that he cured autistic children with his own bone marrow. The claim is patently absurd.) Wakefield's vaccine would be a potential competitor to the MMR and single-shot measles vaccines. The final blow to whatever credibility Wakefield had left was delivered in 2009 when it was discovered that he had fixed his data. Well, one would think it should have been the final blow, but Wakefield will not go away quietly and he still has many advocates. As I write this, there is a Facebook page called Finding a Pox Party in Your Neighborhood run by anti-vaccinationists trying to help others get "natural immunity" to chickenpox for their children. Some parents choose to forgo the chicken pox vaccine for their children and are looking to infect them so they obtain natural immunity. One of the "likes" on the page goes to another Facebook page called Dr. Wakefield's Work Must Continue. One commenter posted the following: "The thing about smart people is that they seem like crazy people to dumb people." True and dumb people seem like dumb people to just about everybody except other dumb people.

In 2004, the British General Medical Council (GMC) began an inquiry into allegations of professional misconduct against Wakefield and two former colleagues (Professor John Walker-Smith and Dr. Simon Murch). Three years later Wakefield was working in Texas at an autism clinic and was still waiting for the GMC hearing to begin. In the meantime, he sued Channel 4, 20-20 Productions, and *London Sunday Times* reporter Brian Deer for libel. He dropped the suits, claiming he had to concentrate on the upcoming GMC inquiry. While being investigated by the GMC, Wakefield was hired as a researcher at a Texas autism clinic with the swell-sounding name of Thoughtful House. He's not licensed to practice medicine in Texas, but tax records show that Wakefield was paid $270,000 for his work at Thoughtful House in 2008 and that Thoughtful House received about $2.4 million in grants and contributions. The GMC finally issued a ruling on Wakefield and two colleagues in January 2010: Wakefield, the Council said, had acted "dishonestly and irresponsibly" in doing his research. According to the BBC:

> The verdict, read out by panel chairman Dr. Surendra Kumar, criticized Dr. Wakefield for the invasive tests, such as spinal taps, that were carried out on children and which were found to be against their best clinical interests....

The GMC also took exception with the way Wakefield gathered blood samples. He paid children £5 for the samples at his son's birthday party. Dr. Kumar said he had acted with "callous disregard for the distress and pain the children might suffer." In February 2010, Wakefield resigned from Thoughtful House. In May 2010, Wakefield and Walker-Smith were found guilty and had their medical licenses revoked. Murch was found not guilty. Wakefield is unrepentant. He has since published a book entitled *Callous Disregard: Autism and Vaccines: The Truth Behind a Tragedy* (2010) with an introduction by actress Jenny McCarthy. (McCarthy denies that she is anti-vaccine. She claims she is pro safe vaccine, yet she seems to think there is no such thing as a safe vaccine.)

Eventually, the vaccine scare created by Wakefield was augmented to focus on the preservative used in the MMR vaccine, thimerosal, which breaks down into ethylmercury when ingested by humans. Thimerosal has been used since the 1930s. It is used in vaccinations to prevent contamination by microbes. (I should say it *was* used since it's been removed from most vaccines because of the scare, not because of any evidence that it is harmful.) The amount of mercury a typical child under two years receives from vaccinations is about 240 micrograms. (A microgram is one millionth of a gram. There are about 28 grams in an ounce.) This small amount is most probably harmless in itself. However, the anti-MMR vaccine lobby led by Lyn Redwood, the mother of an autistic child, claimed that it is not harmless to those who are "especially sensitive" to mercury. The expression "especially sensitive to mercury," however, has no more scientific meaning than "especially sensitive to fluoride." It is an assumption that Redwood and others used in their fight to blame the MMR vaccine for autism in their children and to support their quest to have thimerosal removed from vaccines. They did get their wish regarding the removal of mercury in the MMR vaccine, but even though mercury has not been used in the MMR vaccine for over a decade, the anti-vaccine lobby still refers to the MMR vaccine as "the autism shot," as Jenny McCarthy called it on the Oprah show.

The anti-vaccine lobby is able to call on a few scientists to support their position. Some of these scientists look at the effects of mercury poisoning and compare them to the effects of autism. The parallels are striking, they say. They also note that autism-spectrum disorder diagnoses have increased from 1 in 10,000 in 1978 to 1 in 300 in some U.S. communities in 1999, an increase of 3,330 percent in diagnoses. (The Center for Disease Control surveyed several communities and found about 1 in 150 eight-year-olds is diagnosed with autism spectrum disorder.) While the rate of vaccination has not increased by 3,330 percent, the number of vaccinations a child now receives during the first two years of life has increased, though none of those vaccines now contain thimerosal. In any case, some scientists and many laypeople think that the increase in autism detection parallels the increase in vaccinations and that this correlation indicates a causal connection.

However, correlations are notoriously slippery when it comes to establishing causal connections. The crime rate may have gone down at the same rate that the vaccination rate went up over the past twenty years, but no one (I hope!) would claim that one caused the other just because of a correlation.

Furthermore, nobody doubts the dangers of *mercury* poisoning. But proving that the mercury in thimerosal is not a causal factor in every case of autism in a child who has been vaccinated would be impossible. We have to look at the studies that have been done—all of the studies and not just the ones that fit with our preconceived notions—to determine what is the most reasonable thing to believe about the relationship of vaccines to autism.

Before looking at the science, however, we should be clear about the difference between mercury, methylmercury, and *ethylmercury*, the kind in thimerosal. Fearing ethylmercury molecules because they contain mercury atoms is like fearing salt because it contains chlorine atoms. Ethylmercury is not readily absorbed in humans, but methylmercury is. More important, ethylmercury is excreted much more rapidly than methylmercury and doesn't accumulate in the body. The most common source of methylmercury is not vaccines but contaminated fish. Human exposure to methylmercury can result in long-lasting health effects, especially on fetal development during pregnancy and on neurological development.

Of course, it would be immoral to do a double-blind control study on the effects of thimerosal. It would be unethical to randomly assign children either to get vaccinated or not and then wait to see if a significantly greater number of children develop autism from the vaccination group. The logic behind the control study was described by John Stuart Mill as *the method of difference*. If thimerosal causes autism then we should expect to see a significant difference in outcome between the two groups. Since large numbers of children do not get vaccinated, we can compare the autism rates of those who were vaccinated with those who were not. We don't have to do a randomized, double-blind control study to apply the method of difference. If thimerosal causes autism we would expect to find that the vaccination group has a significantly higher autism rate than the non-vaccination group. In 2002, the *New England Journal of Medicine* published a study done in Denmark that found no significant difference between two groups of children born in Denmark from January 1991 through December 1998. Most of the 537,303 children had been vaccinated (440,655 or 82.0%). Eighteen percent or 96,648 had not been vaccinated. Seven hundred thirty eight children or 0.14% (1 per 728 children) were diagnosed with autistic disorder or autistic-spectrum disorder. The percentage of children in the vaccine group who developed autism did not differ significantly from the percentage of children who developed autism in the non-vaccine group. Had the vaccine group had a significantly higher rate of autism than the non-vaccine group, barring any confounding factors, that would be taken as strong evidence that the correlation is not due to chance and indicates a causal relationship between vaccines and autism. Had the autism rate for the non-vaccine group been, say, 25% higher than in the vaccine group, we would have strong evidence that the vaccine not only does not cause autism but may offer some protection against it.

Another of Mill's methods is the *method of concomitant variation*, which predicts that if two things are causally related then as the cause increases in frequency in a population, the effect should correspondingly increase. And, as the cause decreases in frequency in a population, the effect should correspondingly decrease. Yet, autism rates continued to rise around the world while the use of mercury-based vaccines decreased. A study in Denmark found that as vaccines with thimerosal were removed from the population, the autism rate did not go down as would be expected if there were a causal connection. Instead, the autism rate continued to rise. In January 2008, a study based on data from the California Department of Public Health was published in the *Archives of General Psychiatry* and also found that as vaccines with the mercury-containing additive thimerosal were removed from use, the incidence of autism continued to rise at a steady rate.

In 2004, the Institute of Medicine issued a report from a panel of scientists who had examined scientific studies worldwide and found no convincing evidence that vaccines cause autism. In 2007, a seven-year study of 1,047 children who received mercury-containing vaccines as infants concluded that the data do not support a causal association between early exposure to mercury from thimerosal-containing vaccines and immune globulins and deficits in neuropsychological functioning at the age of seven to ten years. The study, funded by the Center for Disease Control and Prevention, looked for effects such as learning difficulties and developmental delays, but did not assess autism-spectrum disorders. The researchers studied forty-two neuropsychological outcomes. A common error in evaluating such studies is to cherry-pick the data. The study found higher prenatal mercury exposure associated with better performance on one measure of language, on one measure of fine motor coordination, and on one measure of attention and executive functioning. This should not be interpreted to mean that prenatal mercury is good for your child. With forty-two categories to examine, if there is no causal event going on, the statistical odds are that some of the categories will have positive outcomes, some negative, and some neutral.

The Clustering Illusion

Another type of scary story popular in the mass media is the environmental cancer scare. Finding what seems like a statistically unusual number of cancers in a given neighborhood—such as six or seven times greater than the average—seems alarming but it is not rare or unexpected. What percentage of any disease you will find in any given neighborhood depends, in part, on chance and on where you draw the boundaries of the neighborhood. Clusters of cancers that are *seven thousand* times higher than expected—such as the incidence of mesothelioma (a rare form of cancer caused by inhaling asbestos) in Karian, Turkey—are very rare, unexpected, and not likely due to chance. The incidence of thyroid cancer in children near Chernobyl in the Ukraine was *one hundred* times higher after the nuclear power plant disaster in 1986. Such extreme differences as in Turkey and the Ukraine are not expected by chance and are indicative of an environmental cause. But clusters, or seeming clusters, of cancers that seem disproportionate to an area are actually expected by the laws of probability. Those laws do not predict that the same percentage of cancers for a whole state or nation will be found within any given set of geographical boundaries in that state or nation. Epidemiologists refer to the seemingly excessive number of cases in an area as the *clustering illusion.* The intuition that random events occurring in clusters are not really random events gives rise to the illusion.

In 2003, the mother of a child with leukemia in the Sacramento area thought it odd that there were several other people with cancer in her neighborhood. She did a survey and found what appeared to be an excessive number of cancers in the area. I can understand the woman's desire to find something to blame for her child's illness and I can also understand how the average person might be led to think that there must be something in the local environment that is causing the cancers. However, several things should be considered. The woman's boundaries extended to wherever she arbitrarily decided to draw them. She included all kinds of cancers in her survey, not just leukemia. She included people who had lived in the area for various amounts of time, ranging from having been born there to having moved into the area in adulthood. Epidemiologists tried to explain that when you looked at the numbers, they weren't that unusual and didn't warrant investigation into an environmental cause. The data showed no difference in the leukemia rates in her area than in the rest of the Sacramento region. The *Sacramento Bee* took up her cause and eventually there was an official government analysis of the water supply that found no environmental toxins in the water. Still, the *Bee* persisted and identified tungsten found in tree rings as the probable culprit, even though tungsten is not a known human carcinogen and the connection between tungsten in trees and cancer in humans is speculative.

Bad Science: Distant Healing, Telepathy, and Homeopathy

In addition to the clustering illusion, a critical thinker needs to be aware that sometimes the only way we find out about a flaw in a scientific study is when some journalist or other scientist or non-scientist exposes problems with the study. Three cases come to mind: (1) Po Bronson's exposé of the 1998 Sicher-Targ study on distant healing (DH) by prayer; (2) Susan Blackmore's exposé of the work on the ganzfeld ESP experiments done in the lab of Carl Sargent; and (3) James Randi's exposé of the work done on homeopathy in Jacques Benveniste's lab.

The Sicher-Targ study appears to be an extremely well-designed double-blind, controlled study. The original goal of the study was to see if distant healing (DH) could lower the death rate for AIDS patients. Since only one of the patients in the study died, there was nothing to report on that count. The published study, however, claimed that the aim was to measure DH against a long list of AIDS-related symptoms. Sicher went through all the data *after* the study was completed to determine which patients had which symptoms. It was a bit of a fluke that Fred Sicher and Elisabeth Targ's deception was exposed by Po Bronson in *Wired* magazine four years after the study was published. By chance, he happened to discuss the work with the statistician on the project. By that time, Elisabeth Targ had bankrolled the study into another big grant from the National Institutes of Health (NIH). At Targ's urging, the statistician mined the data after the study had been completed. Sicher and Targ reported that a "blind" medical chart review found that subjects who had been prayed for acquired significantly fewer new AIDS-defining illnesses, had lower illness severity, and required significantly fewer doctor visits, and fewer hospitalizations. "These data support the possibility of a DH effect in AIDS and suggest the value of further research," they wrote. However, not only did they change the goal of the study and mine the data after the study was over, they cherry-picked the data to showcase. With two dozen symptoms to chart and only forty patients in the study, it would be expected—even if no causal event was going on—that a few results would be positive, a few negative, and a few would show no effect. It is not appropriate to draw grand generalizations from such small samples. For example, three in the group being prayed for were hospitalized, compared to twelve in the control group. This could be a fluke or it could be related to who had insurance that covered hospitalization. When researchers randomly assign patients to treatment and control groups, if the groups are small they must take care to control for known causal factors. For example, since the original point of the study was to try to reduce mortality by prayer, the researchers should have made sure that they did not randomly assign the older patients who smoke to either group because that would bias the group. Age and smoking are known to be relevant to death rates. Sicher and Targ did not control for smoking in an earlier study on healing prayer and mortality in AIDS patients, which found a significant difference in the treatment group that could have been due to a fluke (there were only 20 people in the entire study) or to smoking. While they controlled for smoking and did other reasonable things to match the two groups in the second study, they did not control for insurance coverage, which would affect days of hospitalization since people with coverage are more likely to go to the hospital than those without. Perhaps they didn't control for insurance because there was no need to since their original intention was to study the effect of prayer on death rates, not length of hospital stays.

Of course, any researcher who didn't report significant findings just because the original study hadn't set out to investigate them would be remiss. The standard format of a scientific report allows such findings to be noted in the abstract or in the discussion section of the report. Thus, it would have been appropriate for the Sicher-Targ report to have noted in the discussion section of their published paper that since only one patient died during their study, it would appear that the new drugs being given AIDS patients as part of their standard therapy (triple-drug antiretroviral therapy) were having a significant effect on longevity. They might even have suggested that their finding warranted further research into the effectiveness of the new drug therapy.

However, in their introductory remarks, the Sicher-Targ report gives the impression that the researchers already knew the new drug therapy would work (making them precognitive!) and that is why they changed the design from the earlier pilot study. In their published paper they claim they never intended to replicate that study—even though that was the basis for getting their funding from NIH. Instead, they wrote:

> ...an important intervening medical factor changed the endpoint in the study design. The pilot study was conducted before the introduction of "triple-drug therapy" (simultaneous use of a

protease inhibitor and at least two antiretroviral drugs), which has been shown to have a significant effect on mortality. [Here, they cite a study, "A controlled trial of two nucleoside analogues plus indinavir in persons with human immunodeficiency virus infection and CD4 cell counts of 200 per cubic millimeter or less," published in September 1997 in the *New England Journal of Medicine, nine months after* their study was supposedly completed! Again, proof of their precognition!] For the replication study (July 1996 through January 1997, shortly after widespread introduction of triple-drug therapy in San Francisco), differences in mortality were not expected and different endpoints were used in the study design.

The above description of why they changed the endpoint is grossly misleading. It was only after they mined the data once the study was completed that they came up with six positive outcomes out of twenty-three possible. In any case, the NIH gave Targ another large grant on the basis of this study and in one of those ironies that defies explanation Elisabeth Targ died at the age of forty-one of a rare form of brain cancer (*glioblastoma multiforme*) before the study—on the effects of prayer on just that rare form of brain cancer—could be completed. Needless to say, she was being prayed for by the entire healing prayer community. Targ was not the first, nor will she be the last, scientist to produce deceitful work in an attempt to prove something she deeply believes in.

In 1990, eight years before the publication of the Sicher/Targ DH study was published, the world of parapsychology was abuzz with the publication in a scientific journal of a meta-analysis by Daryl Bem and Charles Honorton of twenty-eight studies on telepathy, the so-called *ganzfeld studies*. The meta-study indicated something other than chance was going on when subjects, called "receivers," were asked to guess what picture or video other subjects, called "senders," had been looking at in another room while the receiver was concentrating on getting telepathic messages from the sender. (A meta-analysis lumps together the data from several studies and treats the combined data as if it were the result of a single study.) The descriptions of the protocols and procedures used by the six different researchers in different labs looked airtight: they appeared to have eliminated cheating or inadvertent transfer of information by ordinary means (*sensory leakage*). It looked as if something paranormal was going on. The receivers should have guessed correctly only 25% of the time if chance alone were at work, but the hit rate was about 38%. Even skeptics had to agree that such a result was not likely due to chance. Parapsychologist Susan Blackmore visited the lab of Carl Sargent who was getting very good results while Blackmore was not getting any evidence to support the telepathy hypothesis. Sargent's lab had produced about 25% of the entire database (9 of the 28 studies) of the meta-analysis. What she found appalled her. "These experiments, which looked so beautifully designed in print," she said, "were in fact open to fraud or error in several ways, and indeed I detected several errors and failures to follow the protocol while I was there. I concluded that the published papers gave an unfair impression of the experiments and that the results could not be relied upon as evidence for psi [i.e., anything paranormal]" (Blackmore 1987). Blackmore has since left the field of parapsychology behind.

Most of us would not consider visiting a scientist's laboratory to investigate whether he or she was following the procedures and methods described in published papers. In fact, scientific journals do not routinely send out investigative teams to investigate the honesty and integrity of scientists whose papers are accepted for publication. Peer review does not include a review of laboratories. The process is based on trust. It is assumed that scientists will not cheat, though it is known that occasionally a small number of scientists in any field will. Catching cheaters happens occasionally, but it does not usually happen by journals sending out investigative teams to laboratories. Once exception is the case of the journal *Nature,* which did send a team of investigators to examine what was going on in Jacque Benveniste's lab.

In 1988, the eminent science journal *Nature* published an article that seemed to provide strong evidence for homeopathy, whose "medicines" are nothing more than water. Whatever active ingredient the homeopathic remedy maker starts with, a process of dilution is required and the dilution is such that the final potion hasn't a single molecule left of any active ingredient. Benveniste and others claim that water has a "memory" and somehow remembers the presence of the active ingredient, even when no molecules of it remain. In a process that can only be described as magical, homeopathic water's memory is so selective that it remembers only the molecules added by the homeopath and forgets all the others it has come in contact with over the millennia! In any case, the editors at *Nature* knew that the publication would cause an uproar in the scientific community, so they published the article on condition that Benveniste allow an investigative team to visit and examine his lab and procedures during a replication of the work he claimed to have done. The team included an expert in detecting deception, the magician James Randi. The *Nature* investigative team discovered that Benveniste didn't actually do the lab work, which was done by people whose salaries were being paid by a homeopathic pharmaceutical company. The actual lab work turned out to be not exactly as described in the paper that had been published. In fact, the actual processes were so far removed from accepted standards that the investigative committee recommended that *Nature* withdraw the homeopathy article. In any case, when stricter controls and proper protocols were put into place, the people in Benveniste's lab could not replicate their original results.

More Bad Science: Organic Foods and Acupuncture

We should note that even honest studies can be flawed. Sometimes such studies get published in scientific journals and then publicized in the media. I'll limit myself to just two examples here, one study on organic foods and another on acupuncture.

In 2003, Alyson Mitchell, Ph.D., a food scientist at the University of California, Davis, co-authored a paper with the formidable title of "Comparison of the Total Phenolic and Ascorbic Acid Content of Freeze-Dried and Air-Dried Marionberry, Strawberry, and Corn Grown Using Conventional, Organic, and Sustainable Agricultural Practices." The article was published in the *Journal of Agricultural and Food Chemistry*, a peer-reviewed journal of the American Chemical Society. The article got some good press from "green" journalists, who proclaimed that the study showed that organic foods have significantly higher levels of antioxidants than conventional foods. Many people believe that diets rich in antioxidants contribute to significantly lower cancer rates. The data, however, do not support this belief. The data show a *correlation* between such diets and lower disease rates, but the data are mostly observational studies. Studies on antioxidant *supplements* haven't found a correlation between the supplements and various diseases. "Study after study has shown no benefit of antioxidants [taken as supplements] for heart disease, cancer, Parkinson's disease, Alzheimer's disease, or longevity" (Hall 2011). Despite the title of the article, no organic strawberries were tested. The study compared total phenolic metabolites and ascorbic acid in only two crops, marionberries and corn. Both crops were grown organically and conventionally on different farms. The organic berries were grown on land that had been used for growing berries for four years; the conventional berries were grown on land that had been used to grow conventional berries for 21-22 years. The crops were grown on different soil types: the organic soil was "sandy, clay, loam"; the conventional was "sandy, Ritzville loam." The soil for the conventional corn had been used before for wheat; the soil for the organic corn had been used for green beans. The conventional farm used well water; the organic farm used a combination of well and creek water. As you can tell from the title of the article, the metabolites measured were not taken from fresh berries or corn but from samples that had been freeze-dried and air-dried. Though not mentioned in the title, the scientists also compared samples that were simply frozen.

The data provided by the authors in their published study show clearly that there was not enough measurable ascorbic acid (AA) in either of the marionberry samples to compare the organic to the conventional. As already noted, no organic strawberries were studied. There was not enough measurable AA for the freeze-dried or air-dried corn to be compared. So, the only data on AA is for the frozen corn: organic had a value of 3.2 and conventional had a value of 2.1. You can read the study yourself to find out what these numbers represent, but whatever they represent they do not merit the conclusion drawn by the authors of the study: "Levels of AA in organically grown ... samples were consistently higher than the levels for the conventionally grown crops."

The study also compared what it calls "sustainable agricultural practices" to organic and conventional practices. Sustainable practices in this study included the use of synthetic fertilizers. "Our results indicate," the authors write, "that TPs [total phenolics] were highest in the crops grown by sustainable agricultural methods as compared to organic methods." Dr. Mitchell is quoted in the press as saying that their study "helps explain why the level of antioxidants is so much higher in organically grown food." Yet, her study clearly states that the evidence for this claim is anecdotal. In fact, the authors write of the comparative studies that have been done:

> These data demonstrate inconsistent differences in the nutritional quality of conventionally and organically produced vegetables with the exception of nitrate and ascorbic acid (AA) in vegetables.

Another study led by Mitchell claims that organic tomatoes have "statistically higher levels ($P < 0.05$) of quercetin and kaempferol aglycones" than conventional tomatoes. The increase of these flavonoids corresponds "with reduced manure application rates once soils in the organic systems had reached equilibrium levels of organic matter." In fact, the study suggests that it is the nitrogen "in the organic and conventional systems that most strongly influence these differences." The authors suggest that "overfertilization (conventional or organic) might reduce health benefits from tomatoes." The argument is that the flavonoids are a protective response by the plants and one of the things they respond to is the amount of nitrogen in the soil. In any case, the thrust of these and similar studies is that both organic and conventional crops can be manipulated to yield higher levels of antioxidants. At least one study has found "organic food products have a higher total antioxidant activity and bioactivity than the conventional foods." That study, however, involved only ten Italian men.

The desire by some scientists to demonstrate the superiority of organic foods is matched by the desire of others to prove that acupuncture should be considered as a viable treatment for an assortment of ills and behavioral problems. For a review of many of the scientific studies done on acupuncture, I recommend R. Barker Bausell's *Snake Oil Science: The Truth about Complementary and Alternative Medicine*. The evidence strongly supports the view that acupuncture is a placebo therapy. Thus, one of the key elements of a well-designed acupuncture study is that it control for the placebo effect.

One study not covered by Bausell (because it was published after his book came out) is typical of these studies. I became aware of the study by reading an article on the *New Scientist* magazine website on December 20, 2007, with the headline: **Acupuncture Relieves Cancer Chemotherapy Fatigue**. Chemotherapy wipes people out, and any treatment that would provide a boost in energy would be welcome. If researchers were doing disinterested science, they would do a large study (at least twenty-five in each group), and it would be double-blind, randomized, placebo-controlled, have a low attrition rate, and be published in a high quality scientific journal. (See Bausell 2007: 104). The acupuncture study touted by *New Scientist* had forty-seven participants and was described by the researchers as "a randomised placebo-controlled trial." The patients were randomly assigned to one of three groups to receive either acupuncture or acupressure—placing physical pressure on acupuncture points with hands or objects—or sham acupressure.

The acupuncture group received six 20-minute sessions over a period of three weeks. The acupoints "were selected for their supposed propensity to boost energy levels and reduce fatigue." The acupressure group administered their own therapy. They were taught to massage the same acupoints for one minute a day for two weeks. The members of the sham acupressure group also administered their own therapy, but they were given different points to massage. Some key ingredients for a placebo-controlled study on acupuncture seem missing: there is no sham acupuncture group and there is no administration of the therapy by the healer in a clinical setting for the two acupressure groups. The way this study was conducted meant that compliance with the acupuncture group was known and was likely to be high, whereas compliance with the acupressure groups would have to rely on self-reporting. In fact, even though this is a rather small study, one would predict—based on what we know about the placebo effect—that the difference in method of delivery of the treatment would lead to the acupuncture group reporting the best results. Another defect in the study is that the acupressure groups applied their therapy a minute a day for two weeks (28 minutes of self-treatment), while the acupuncture group received its therapy for three weeks (two 20-minute sessions per week, for a total of 120 minutes of therapy in a clinical setting). The difference in delivery of treatment by a healer in a clinical setting versus self-administration could account for any difference in effect, as could the difference in duration of the treatments.

The results were that "patients in the acupuncture group reported a 36% improvement in fatigue levels, whilst those in the acupressure group improved by 19%. Those in the sham acupressure group reported a 0.6% improvement." Without knowing how patients in a sham acupuncture group would have responded, we don't know what the 36% improvement reported means. Acupuncture worked better than acupressure, but that may be because the patients had more reason to believe in the effectiveness of the acupuncture therapy and more reason to expect good results. The large difference between the two acupressure groups is interesting, however. It could be an artifact of the small size of the samples, of the dropout rate (not mentioned in the article), or of the way improvement was measured. It could indicate that traditional trigger points of acupressure are more effective than non-trigger points. Whatever this study indicates, there is little justification for claiming acupuncture relieves cancer chemotherapy fatigue. The study was published in the peer reviewed journal *Complementary Therapies in Medicine* (Volume 15, Issue 4, Pages 228-237). It's called "The management of cancer-related fatigue after chemotherapy with acupuncture and acupressure: A randomised controlled trial." It was done at Manchester's Christie Hospital by Alexander Molassiotis et al.

Many people think that acupuncture can't be explained by the placebo effect because it (and other therapies whose effectiveness is attributed to the placebo effect) has a real, measurable physiological component. However, Antonella Pollo and colleagues have shown that the placebo effect can stimulate the opioid system. Their work was published in the Journal *Pain* (July 2001) as "Response expectancies in placebo analgesia and their clinical relevance." The following is from their abstract:

Thoracotomized patients were treated with buprenorphine [a powerful pain reliever] on request for 3 consecutive days, together with a basal intravenous infusion of saline solution. However, the symbolic meaning of this basal infusion was changed in three different groups of patients. The first group was told nothing about any analgesic effect (natural history). The second group was told that the basal infusion was either a powerful painkiller or a placebo (classic double-blind administration). The third group was told that the basal infusion was a potent painkiller (deceptive administration). Therefore, whereas the analgesic treatment was exactly the same in the three groups, the verbal instructions about the basal infusion differed. The placebo effect of the saline basal infusion was measured by recording the doses of buprenorphine requested over the three-days treatment. We found that the double-blind group showed a reduction of buprenorphine requests compared to the natural history group. However, this reduction was even larger in the deceptive administration group. Overall, after 3 days of placebo infusion, the first group received 11.55 mg of buprenorphine, the second group 9.15 mg, and the third group 7.65 mg. Despite these dose differences, analgesia was the same in the three groups. These results indicate that different verbal instructions about certain and uncertain expectations of analgesia produce different placebo analgesic effects, which in turn trigger a dramatic change of behaviour leading to a significant reduction of opioid intake.

The patients who thought their IVs contained a powerful pain reliever required 34% less of the analgesic than the patients who weren't told anything about their IVs and 16% less than the patients who were told their IVs could contain either a powerful pain killer or a placebo. Each group got exactly the same amount of pain killer but their requests for the analgesic differed dramatically. The only significant difference among the three groups was the set of verbal instructions about the basal infusion.

Critical Thinking about Scientific Studies in the Media

So, what is a critical thinker to do? If we can't assume that an article published in a peer-reviewed journal is trustworthy, how are we to assess scientific papers reported on by the media? If we can't trust scientists to do what they say they are doing in their labs, then what value can we put on the conclusions of their studies? How is the layperson to know whether a study has been properly designed or controlled? There are a few rules we can follow. *We should be skeptical of single studies.* That should be the case even if we have every reason to trust the lab. Until the work is replicated under a variety of conditions, we should suspend judgment. *We should be skeptical of studies that have no controls. We should not treat observational studies as proving causal connections, no matter how strong the correlations that are discovered. We should trust consensus science, even though the majority is not always right.* Scientists will get it wrong occasionally, as they did with ulcers. For many years, the consensus among scientists and doctors was that stomach ulcers were caused by stress. We now know that stomach ulcers are caused by bacteria. Science is self-correcting. Scientists will eventually get it right. And, the fact is, that despite the occasional cases of deception like that of Wakefield and Targ, there really is no compelling evidence that in general scientists are dishonest and out to trick us.

Currently (October 2011), there is a debate about the overuse of magnetic resonance imagery (MRI) scans in sports medicine. Many orthopedic surgeons routinely order such scans for injured athletes. The scans almost always find some abnormality, and surgery is often recommended on the basis of the scan. Dr. James Andrews, a highly respected sports medicine orthopedist, knew that scans almost always find something abnormal, although most abnormalities are of no consequence. He scanned the shoulders of thirty-one perfectly healthy professional baseball pitchers and found abnormal shoulder cartilage in 90 percent of them and abnormal rotator cuff tendons in 87 percent. Dr. Andrews and other eminent sports medicine specialists are speaking out about the overuse of magnetic resonance imaging and the concomitant overuse of surgery for athletic ailments. Less controversial are the full body scans being offered people who are healthy but want to know if there might be something wrong with them that they don't know about. There is a consensus among physicians that full body CT (computed tomography) scans are dangerous on three counts. One, the abnormalities they discover may be of no medical consequence except to scare hell out of the patient. Two, many more people will die of cancer caused by the high radiation doses used in CT scans than will be saved by discovery of a hitherto unknown cancer in the patient. And three, figuring out whether incidental findings are significant can be expensive and time consuming, not to mention the danger of further invasive tests that might even include exploratory surgery. A normal CT scan of the chest is the equivalent of about 100 chest X-rays, but some scanners are giving the equivalent of 440 conventional X-rays.

Most of the time, the consensus position is based on solid evidence and reasoning. If you want infallibility, you're on the wrong planet. One last rule: *don't give too much value to small studies.* Even if the study is a double-blind, randomized study, if it only includes thirteen people, the fact that two of them have eye problems during the study should not be taken as evidence that their eye problems had anything to do with the study. This example is not purely fictional. A doctor (Ralph G. Walton, M.D., Department of Psychiatry, Western Reserve Care System) with a known bias against aspartame had a study halted after two subjects, one each in the placebo and the control groups, developed eye problems. Nothing significant should be concluded from the fact that two of thirteen people, one in the aspartame and one in the placebo group, developed eye problems. These are most likely coincidental and chance occurrences. Only eleven out of forty subjects completed the three-week study. Three dropouts said they felt like they had been poisoned, even though nobody has ever reported poisoning symptoms from aspartame. Two subjects completed only part of the study but were included in the data. Despite all these problems, Dr. Walton published his incomplete study in the peer-reviewed *Journal of Biological Psychiatry* in 1993 and stated: "We conclude that individuals with mood disorders are particularly sensitive to this artificial sweetener and its use in this population should be discouraged." To make matters worse, Walton counted the individual with eye-problems in the placebo group as if it had occurred in the aspartame group on the grounds that this was a crossover study and the one in the control group had been in the aspartame group before crossing over to the placebo group. The design of the study was such that after seven days in either group, the participants would go through a three-day "washout" period where they would take neither aspartame nor placebo. (The study was blinded, so the participants never knew which substance they were ingesting.) The person who had eye-problems while in the placebo group had not taken any aspartame for six days before he started having eye issues. The "washout" period of three days must have been chosen for a reason. Otherwise, this study could do nothing but support the bias of the researcher, since any problems that occurred to the placebo group could always be alleged to have been caused by delay from when the participant was in the aspartame group. Finally, it should be noted that the conclusion Dr. Walton drew was about people with "mood disorders" in general, yet his study included only people with a history of depression or no history of depression. There is no reason to assume that people with bipolar disorder, for example, will react the same way as those with unipolar depression. Generalizations from sample studies should not extend to broader samples than the one tested. The moral of the story is *beware of studies done by scientists with axes to grind*. Dr. Walton's bias led him to attract subjects who shared his bias (*experimenter bias*) and draw conclusions based on his expectations rather than on the evidence (*expectancy bias*). Another moral is that when you are reading a newspaper or magazine account of a scientific study look for the following pieces of information: *How was the study done? How many were in the study? What controls were used? How many dropped out of the study? Who paid for the study? Does the conclusion extend to a group that was not in the sample? Does the researcher indicate any bias?* Dr. Walton states in his published paper that he has clinical experience of patients with affective disorders who appear particularly prone to adverse reactions to aspartame. He recruited participants from the place he worked; the recruits probably knew of his position on aspartame. He recruited forty people to participate but only eleven finished the three-week trial. Clearly, neither the scientist nor the subjects appear to be free from bias.

The critical thinker looks for evidence of bias in the research and avoids cherry-picking scientific studies: don't accept a study just because it fits with your beliefs and don't reject a study just because it conflicts with your beliefs. Some people who already believe that aspartame is unhealthy might be tempted to cite Dr. Walton's study as supporting their view. After all, it's a "scientific" study and was published in a peer-reviewed journal! Avoid this temptation. Supporting your viewpoint with poor scientific studies will make you look foolish and could be harmful to your health. Finding good information is sometimes difficult or inconvenient. Many people now go to "Google University" to do their medical research. If you use the Internet as your main source of medical information, please consult Science-Based Medicine (www.sciencebasedmedicine.org/) and Quackwatch (www.quackwatch.com) as part of your research.

Another problem the critical thinker has to contend with is scientific data that seem to support a hypothesis but are actually irrelevant to it. For example, the Redwoods' concern about mercury being tied to autism began when they read a report in 1999 from the Food and Drug Administration (FDA) that said babies who receive multiple doses of vaccines with thimerosal may be exposed to more mercury than recommended by federal guidelines. The Environmental Protection Agency (EPA) says "five parts per million is diagnostic for mercury toxicity." When he was four or five, the Redwoods' son had mercury levels in his hair of 4.8 parts per million, according to his mother. However, the amount of mercury in hair does not reflect the concentration in the rest of the body. According to Dr. Robert Baratz, "analyzing hair for mercury is a waste of time and money and cannot be used to diagnose mercury poisoning. A competent practitioner would easily know this." Maybe. But the competent layperson wouldn't be likely to know this and thus could be easily duped into thinking he or she has a problem when none exists.

We have to be careful that we don't go overboard in our skepticism and reject a study simply because of *possible* bias. This is what seems to have happened to Mark Blaxill of SAFE MINDS (Sensible Action for Ending Mercury-induced Neurological Disorders). He accused the authors of the Danish study that found that autism was increasing while thimerosal was being removed from vaccines of manipulating "the incidence of autism in an attempt to clear thimerosal-containing vaccines of any role in the etiology of the disease." Why would these scientists intentionally manipulate data to exonerate thimerosal? Because, said Blaxill, pediatricians, the ones who read *Pediatrics* (where the Danish study was published), administer vaccines and he thought they wanted to stop the movement to eliminate thimerosal from vaccines. Blaxill believes that it is damning that two of the authors of the study work for the Danish manufacturer of thimerosal vaccines and *Pediatrics* didn't mention this. Nor did the journal mention that it gets advertising revenue from manufacturers of vaccines. However, it is appropriate for someone who works for a manufacturer of a product that has been claimed to be harmful to be involved in a study on the effects of that drug. I could understand Blaxill's complaint if the researchers had found that as thimerosal decreased so did autism *and* then they had refused to publish the study. Also, the fact that manufacturers of vaccines advertise in *Pediatrics* seems pretty lame as a reason for rejecting the study outright, though it might raise a red flag. In any case, it seems obvious that pediatricians would continue to give vaccinations whether they contained thimerosal or not.

We'll conclude this chapter with a review of other ways scientists can mislead us and with some general comments about how scientists establish causality.

Misusing Graphics, Statistics, and Meta-analysis

Sometimes scientists produce confusing graphical displays of data to serve their own interests. Dean Radin does this in his *Conscious Universe: The Scientific Truth of Psychic Phenomena* when he graphically displays the data for a meta-analysis of aspirin studies. As noted above, a meta-analysis lumps together the data from several similar studies and uses statistical analyses of the data as if they came from one large study. Critics mockingly describe a meta-analysis as one where you lump together ten studies that didn't find anything significant and magically produce a significant result that defies the odds by a trillion to one. An example of this kind of misuse of statistics can be found in Radin's book *Entangled Minds* where Radin reports on four tests of presentiment he had done that produced mixed results. Two of his studies produced statistics that favored the presentiment hypothesis and two didn't. He did a meta-analysis of the four studies and found statistical significance. The concept of *statistical significance* can be misleading. It signifies nothing more than that the results are *not likely due to chance*. It does not necessarily signify that the results are important, nor does it signify that a causal connection has been established. In any case, from his small studies that produced mixed results, Radin claimed that a meta-analysis showed that the odds against chance of his results were 125,000 to 1 (pp.166-168).

Before discussing Radin's misleading use of the visual representation of data, I'll review how he has abused meta-analysis. For Radin, meta-analysis has provided the scientific proof of the reality of ESP and psychokinesis (together they are called *psi* by parapsychologists). Even with meta-analysis, however, parapsychologists are able to find only very small differences between what one might expect by chance and what one might expect if a paranormal power were at work. For example, in 1986 Robert Jahn, Brenda Dunne, and Roger Nelson reported on millions of trials with thirty-three subjects over seven years trying to use their minds to override random number generators (RNG). Think of the RNG as producing zeros and ones. Over the long haul, the laws of probability predict that in a truly random sequence, there should be 50% of each produced. The subjects in these experiments tried to use their minds to produce more zeros (or ones, depending on the assignment). In short, the "operators," as they were called by those who did the experiments at Princeton University, were trying to do what many drivers try to do when stopped at a red light and will the light to turn green. (It is fascinating how attractive this idea is. Being able to make things happen by willing them seems to be an innate desire.)

In 1987, Radin and Nelson did a meta-analysis of all RNG experiments done between 1959 and 1987 and found that they produced odds against chance beyond a trillion to one (Radin 1997: 140). This sounds impressive, but as Radin says "in terms of a 50% hit rate, the overall experimental effect, calculated per study, was *about* 51 percent, where 50 percent would be expected by chance" (p. 141, emphasis added). A couple of sentences later, Radin gives a more precise rendering of "about 51 percent" by noting that the overall effect was "just under 51 percent." Similar results were found with experiments where people tried to use their minds to affect the outcome of rolls of the dice, according to Radin. When Nelson did his own analysis of all the Princeton data (1,262 experiments involving 108 people), he found similar results to the earlier RNG studies but "with odds against chance of four thousand to one" (Radin 1997: 143). Nelson also claimed that there were no "star" performers. However, according to another analysis by psychologist Ray Hyman, "the percentage of hits in the intended direction was only 50.02%" in the Princeton studies (Hyman 1989: 152). Either way, the difference between chance expectation and the actual outcome for millions of trials is tiny. These data should remind us that statistical significance does not imply importance.

Radin disagrees. He thinks that small differences can be important. To prove it, he produces some graphs that try to illustrate that scientists in parapsychology are not different from scientists in medicine who treat small differences as having big effects. Radin depicts a meta-analysis of twenty-five aspirin studies. He depicts the data using data point estimates within a 99% confidence interval, which is misleading because it completely hides the significance and information conveyed by the data of those twenty-five studies. (The 99% confidence interval means a statistical formula was used that makes it highly unlikely [a one in one-hundred chance] that the results are a fluke.) His graphic is shown in figure 6.

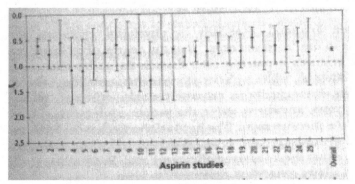

Figure 6

The x-axis lists 25 studies and is labeled "Aspirin studies." The y-axis, not labeled in the graphic above, is labeled the "ratio" on Radin's graph and runs from 2.5 at the bottom to 0.0 at the top. He says that a value of 1.0 means that the treatment was no better than the control in reducing heart attacks. We'll take his word for it. In any case, his visual does not make it clear that there were 22,071 physician-subjects in those studies and that there were 104 heart attacks among the 11,037 subjects in the aspirin group and 189 heart attacks among the 11,034 subjects in the placebo group. His graphic display does not depict that the odds of this difference being due to chance are on the order of 100,000 to 1. His visual doesn't convey that there were 44% fewer heart attacks in the aspirin group or even that there were 85 fewer heart attacks in the aspirin group. It is true that only 0.94% of the aspirin group had heart attacks, while 1.71% of the placebo group had heart attacks. This may not seem like much of a difference, but since more than half a million people die each year of heart disease and 80% of heart disease deaths in people under 65 are due to first heart attacks, this means that an aspirin a day *could* extend the lives of tens of thousands of people each year.

The way Radin depicts the data only five of the twenty-five studies showed positive results (i.e., at the 99% confidence interval they had both their extremes on the positive—upper—side of the graph line across from 1.0). One wonders why he portrays this data at the 99% confidence interval, instead of at the usual 95%. Perhaps it had something to do with the fact that at the 95% level most of these studies would have their extremes on the positive side of the graph, thereby nullifying his point that a bunch of studies that are unimpressive can be impressive when lumped together. While Radin is correct in noting that "the aspirin effect was declared to be 'real' based on the combined results of all studies," he is wrong to imply that in the case of the aspirin studies most of them were unimpressive or didn't show a positive effect. Thus, his conclusion that "this is exactly what meta-analysis has done for psi experiments" is a gross exaggeration.

Not only is the visual display of this data misleading, it is disinformative. It does not reveal the dramatic effect of aspirin on preventing first heart attacks. Furthermore, Radin doesn't even mention that the aspirin studies were stopped in midstream because the data were so convincing that it would have been unethical to have continued the study and not offer the benefit of aspirin to those in the placebo group, as well as to make known the findings to the general public. Thus, had the studies been carried out to completion, it is likely that at the 95% confidence interval, the vast majority of the studies would have supported the positive effect of aspirin. In fact, later studies have replicated the earlier studies. In 2003, a meta-analysis by researchers at Mount Sinai Medical Center & Miami Heart Institute of five major randomized clinical trials on aspirin with 55,580 participants, including 11,466 women, found that aspirin reduces the risk of a first heart attack by 32 percent and that aspirin reduces the combined risk of heart attack, stroke, and vascular death by 15 percent.

Radin tried to make the aspirin study look like the psi studies: take a bunch of losers, lump them together and declare them a winner. It's as if you could take 20 years of batting averages below .250 and lump them together and somehow magically produce a lifetime average of .357.

Meta-analysis is certainly a useful tool in the right hands, but it can be easily abused in fields like parapsychology where the studies are very unequal in quality. Bias can creep in when setting the criteria by which the analyst decides which studies to include and exclude. Many negative studies don't get submitted or published if they are submitted, so they are not considered in the meta-analysis. This is known as the *file-drawer effect*. In addition, there is the problem of *publication bias*. Journals tend to publish studies with positive outcomes and some scientists don't try to publish results that don't jibe with their biases.

Common Errors in Causal Reasoning

We'll conclude this foray into the world of critical thinking about science publications with some brief comments about common errors in causal reasoning. We've already mentioned one of the most common causal fallacies: the post hoc fallacy. Just because one thing occurs after another doesn't mean the first thing caused the second. It is true that every causal event involves something that occurs first (the cause) and something that happens later (the effect), but many things happen one after the other that are *not* causally related. It's also true that causal relationships are correlational. If smoking causes lung cancer then if there is an increase in the number of heavy smokers in a stable population there should be a rise in the number of people with lung cancer. But many things are correlated that are not causally related. Further complicating matters is that when two things are correlated *and* causally related, the correlation doesn't tell us which one is the cause and which the effect; in fact both could be effects of a third factor. For example, a study published in the *Journal of the American Medical Association* in October 2011 found that men who took high daily doses of vitamin E (400 units) for about five years had 17% more cases of prostate cancer than a control group. This study found a correlation between taking large doses—the daily recommended dose is about 23 units—and prostate cancer. It did not find a causal connection between vitamin E and prostate cancer. Yet, a typical report in the media on this study said things like "There is more evidence that taking vitamin E pills can be risky," "Vitamin E pills may raise risk of prostate cancer," and "Men randomly assigned to take a 400-unit capsule of vitamin E every day for about five years were 17 percent more likely to get prostate cancer than those given dummy pills." The main conclusion one should draw from this study—which was an attempt to replicate previous, less rigorous, studies that found vitamin E supplements protected men against prostate cancer—is that the earlier studies were wrong.

There have been studies that have found that people who go to church are healthier than those who don't and that people who take vitamins have a higher mortality rate than people who don't take vitamin supplements. Going to church may not be good for your health, however. It may be that if you're not healthy, you can't go to church. And many people who take vitamin supplements may do so because they're not healthy and hope the vitamins will give them a boost. It's not necessarily the case that the vitamins are the cause of their higher mortality rate.

Causal fallacies are just one of many kinds of errors the human mind is naturally drawn to. How many kinds of fallacies are there? Let me count the ways....in the next chapter.

SOURCES FOR CHAPTER SEVEN 274

VIII
The Fallacy-Driven Life

"I intend to open this country up to democracy and anyone who is against that I will jail, I will crush."—João Baptista Figueiredo, President of Brazil (1979)

"...sometimes I've believed as many as six impossible things before breakfast."—the White Queen in *Alice in Wonderland*

Fallacies are errors in reasoning. They drive the thought-engine of most people most of the time. We did not evolve to seek truth, beauty, and goodness. We evolved to survive and mate, though we can surely take pride in our noble reason.

> What a piece of work is a man!
> How noble in reason! How infinite in faculty!
> In form and moving how express and admirable!
> In action how like an angel!
> In apprehension how like a god!
> The beauty of the world! The paragon of animals!

Hamlet mocks our species. In fact, man is also an irrational animal, driven by his needs, fears, and wants, following logic or reason if it suits him. Our natural way of thinking, of making judgments, of identifying causal connections is to jump to conclusions on flimsy evidence. (One wag noted that for some people jumping to conclusions is their only form of exercise.)

Critical thinking is unnatural. Following our feelings and emotions is more likely to motivate our behavior than well-reasoned arguments. We are as likely to be persuaded by irrelevant appeals as by relevant ones, and are more likely to produce slanted, selective, biased, one-sided, incomplete arguments than well-reasoned, fair-minded, reflective, accurate, complete arguments. We often prefer attacks on a person's motives to attacks on a person's reasons. We make assumptions that aren't warranted, create straw man arguments out of fragments of

opposing viewpoints, offer up false dilemmas, and draw
conclusions hastily. It's amazing we've made so much progress!

To Be or Not to Be Irrelevant

On a quiet Tuesday morning in October in the Sacramento suburb of Elk Grove, California, F.B.I. agents knocked on Michael Lee Braun's door and arrested the nuclear engineer for mailing threatening letters to President George W. Bush and Vice President Dick Cheney. Braun had been under scrutiny for five years, suspected of sending baking soda in envelopes marked inside with the word "anthrax" to those on his enemies list. Braun's attorney was asked by a local newspaper reporter to comment on his client's arrest. Did his client deny the charges? No. Did the attorney claim that the F.B.I. had arrested the wrong man? No. Did the attorney say anything *relevant* to the arrest or the charges? No. He did what many people in the public eye do when asked a direct question about their own wrongdoing or that of someone dear to them: he questioned the *motives* of those asking the question and making the accusations. "Why did they wait until the president showed up to arrest him?" asked Mr. Cozens, Braun's lawyer, referring to the fact that President Bush was in the area campaigning for Republicans. Cozens continued, "It seems odd to me when I read the complaint that they had waited so long to do this."

What seems odd is that this rather common cheap trick of questioning the motives of people when something else is at issue passes by regularly without notice or comment. In America it happens on an almost daily basis that some politician who is accused of wrongdoing responds by questioning the motives of those reporting on the matter. "This whole thing is politically motivated," the accused cries out. How refreshing it would be to hear a journalist respond to this oft-heard bellow by responding: *So what? Answer the question, please.*

When Rose Bird was Chief Justice of the California Supreme Court, her court was assailed time and again for making decisions that overturned death penalty convictions. The attacks were not backed up by appeals to the law, the Constitution, precedent, or anything that might have been relevant to making a proper judicial decision. Instead, the court's critics found it sufficient to attack the motives of those judges who voted for overturning death penalty cases. These judges, the critics said, are philosophically opposed to the death penalty. The proper response to such a complaint is *so what?* Their motives are irrelevant to the reasons they gave in their rulings. Finding fault with the *reasons* given for a decision is the proper way to attack a decision. Whether the decision issued from a lousy childhood in some previous lifetime or from a philosophical position developed over this lifetime is irrelevant to the issue of the cogency of the decision.

During the 2011 campaign by Republicans to pick their 2012 presidential candidate, several women accused Herman Cain of having sexually harassed them years ago. His response was to claim the allegations were "the work of political insiders trying to prevent a businessman from being elected U.S. president." That might have been true, but it's irrelevant to whether he harassed the women.

Attacking motives rather than reasons is common in the many attacks made on critics of "alternative" medical treatments. Such critics are often met with the response: "You must be in the pocket of Big Pharma." I would be one of the first to admit that Big Pharma deserves criticism. But even that—providing direct criticism of Big Pharma—seems to be too much to ask of those defending themselves against critics by asserting that the critic must be a shill for some pharmaceutical firm. But rather than taking apart Big Pharma's actions point by point, we usually hear the opponent of Big Pharma claim that *it's just playing politics* or *it's in the pocket of the AMA (American Medical Association)*. Critics of conspiracy "theorists" are met in a similar fashion.

One critic of my arguments against the claim that the Bush administration was behind the terrorist attacks on September 11, 2001, had nothing to say about my argument (which can be found at <skepdic.com/refuge/bunk27.html>) but a lot to say about me. For example, he wrote: "you're probably an old guy and as we get older the brain just doesn't wanna have to deal with reality." Both claims may be true but neither of them is relevant to refuting my argument. This same fellow also wrote: "your views support a system that is completely corrupt because you have all your retirement money invested in that same system." Instead of providing an argument demonstrating that I shouldn't trust any government report or claim on the issue of 9/11, he wrote that my "'don't wanna question my government' view on 9/11 is scary quite frankly." This last assessment borders on a straw man attack, where one distorts another's position to make it easier to attack. This fellow seems to be suggesting that my real argument was that we should trust our government whenever it tells us something. Such a caricature of my argument passes from being an attack on me into being an attack on a position I don't hold and did not defend.

Good refutations of arguments try to undermine the accuracy, relevance, fairness, completeness, and sufficiency of reasons given to support a conclusion. One of the more common tactics of those who can't provide a good refutation of an argument is to divert attention away from the argument by calling attention to something about the person who made the argument. Rather than criticize a person's premises or reasoning, one asserts something about the person's character, associations, occupation, hobbies, motives, mental health, likes, or dislikes. Attacking motives is a favorite of lazy thinkers. It doesn't take much skill or time to ridicule or belittle a person. It's often hard work to seriously examine an opposing viewpoint. Attacking motives is a tactic of the clever manipulator of crowds, the experienced demagogue who knows how to play on the emotions of people and seduce them into transferring their attitude of disapproval for a person to disagreement with that person's position.

Criticizing a person's motives rather than his arguments is called an *ad hominem* (to the man) fallacy. Attacking a person, rather than the person's position or argument, is usually easier as well as psychologically more satisfying to those who divide the world into two classes of people—those who agree with them and are therefore good and right, and those who disagree with them and are therefore evil and wrong. It seems that many of us make an association between what is perceived as being shady or questionable in a person and that person's argument being shady or questionable. But a minute's reflection reveals that a good person can make a bad argument and a bad person can make a good argument. The argument stands or falls depending on the evidence presented for the conclusion, not on the character or motives of the arguer. Even so, sometimes the easiest way to get someone to dislike an argument is to get them to dislike the arguer. That's why so many people use the Hitler card, the liberal card, the socialist card, the Tea Party card, or the atheist card. They know that by associating an opponent's position with someone or some group their audience will consider evil, they can persuade many to reject their opponent's arguments without even considering the evidence presented in those arguments. This also saves the lazy person from having to prove that his position is the better of the two.

One of the more common *ad hominem* tactics might be called *using the money card*. Instead of providing a refutation of a position you claim is in error, you accuse the one who has made an argument for the position of making money from it. Some opponents of vaccinations against diseases such as measles and whooping cough don't challenge the reasons for giving such vaccinations; instead, they point out that pharmaceutical companies, medical clinics, and physicians make money from the sale of vaccinations and from providing them to their patients. So what? Whether anyone makes money from the sale of vaccines is irrelevant to whether there are good reasons for those vaccines. Sometimes the charge of making money from a position is false. The fallacy in the *ad hominem* is due to the *irrelevant* nature of the appeal made, not to its falsity. If what is said about the person is false in addition to being irrelevant, two fallacies are committed, *false premise* and *irrelevant premise*.

When the irrelevant claims are negative, the *ad hominem* fallacy is called *poisoning the well*. This ploy is a not-so-clever but effective tactic. Here are some examples. "Now that we've heard from the thoughtful group, is there anything any of you bearded anarchists have to say?" "Anyone who opposes the President's deployment of troops to Saudi Arabia is a traitor! They should be put in a cell with Hanoi Jane Fonda. But, this is America, so go ahead and have your say." "Before you begin spouting your liberal propaganda, let me just say...." "This just in from the femi-nazis."

"And here's one from the animal rights wackos." "Atheism is just a religious cult of egomaniacs."

Demeaning a person rather than critiquing his argument gives one a false sense of license to avoid producing any evidence of your own while giving the illusion of providing a rebuttal. It also creates the false impression that the position you take is held in good faith, while the position you oppose is held by corrupt or compromised people.

The *ad hominem* fallacy is often confused with the legitimate provision of evidence that a person is not to be trusted. Calling into question the reliability of a witness is relevant when the issue is whether to trust the witness. It is irrelevant, however, to call into question the reliability or morality or anything else about a person when the issue is whether that person's *reasons* for making a claim are good enough reasons to support the claim.

Many who are duped by the *ad hominem* appeal will also be duped by what's sometimes called the *argumentum ad ignorantiam* (argument to ignorance). That fallacy involves reasoning that because one of two opposing positions hasn't been shown to be true then the other one must be true. But, just because somebody can't prove that Thor doesn't exist, doesn't mean that Thor exists. Even if you agree to reject my opponent's position because I've convinced you that he is an idiot (rather than that his argument is faulty), that doesn't mean that my argument is a good one or that my position is the most reasonable one given all the evidence. There might be a third position that is even better than either of ours. If I can get you to falsely assume that there are only two reasonable positions on an issue, and I can get you to reject my opponent's argument, I might sucker you into accepting mine without providing any evidence that mine is cogent. Many a person has been duped by the *false dichotomy*, also known as the *false dilemma* or the *black and white fallacy*: assuming there are only two positions on an issue when there are actually more than two.

There are many other irrelevant appeals that are as popular as the *ad hominem*. One of the most popular irrelevant appeals is the *appeal to popularity* itself. Even though it is obviously irrelevant to the truth of a claim that many people accept it, this fact hasn't deterred too many of us from trying to get support for an idea by pointing out how many people agree with us. The fact that billions of Chinese use acupuncture to unblock a subtle energy—assuming they do—does not mean acupuncture unblocks subtle energy. Millions of people spend billions of dollars every year on worthless "health" products, especially multivitamins, that are promoted by advertisements that tout how many people use the products. Fifty million people can't be wrong! Right? No. Actually, 50 million people can be wrong. Think of all the millions of people who believed the light from the stars came in through holes in a cosmic sky dome. How many people once believed in bloodletting as a way to get rid of disease? It would be good if 50 million people believed in the effectiveness of some medical treatment because the scientific evidence provides strong support for its effectiveness. It is not so good if one person believes a treatment is effective only because 50 million other people believe it is.

The irrelevant appeal to popularity is also known as the *ad populum* fallacy, the *bandwagon fallacy, the appeal to the mob,* and *the democratic fallacy*. This fallacy is seductive because it appeals to our desire to belong and to conform, to be part of the in-crowd. Its appeal is enhanced by the illusion that in-crowds give to our sense of safety and security. It is a common appeal in advertising and politics. A clever manipulator of the masses will try to seduce those who blithely assume that the majority is always right. Also seduced by this appeal are the insecure, who may be made to feel guilty if they oppose the majority or feel strong by joining forces with large numbers of other uncritical thinkers.

Closely related to the irrelevant appeal to popularity are the *irrelevant appeal to tradition* and the *irrelevant appeal to authority*. The irrelevant appeal to tradition is a fallacy in reasoning in which one argues that a practice or a belief is justified simply because it has a long and established history. For example, the U.S. Supreme Court has ruled that requiring prayer in government institutions is unconstitutional. Those who objected solely on the grounds that public schools have had compulsory prayer for many years, and so the Court was wrong to end this practice, were making an irrelevant appeal to tradition. Those who argue against same-sex marriage in America solely on the grounds that marriage has always been between a man and a woman in this country are making an irrelevant appeal to tradition.

Stare decisis, or the custom of many legal systems that requires respect for judicial precedents, should not be confused with the irrelevant appeal to tradition. Judicial decisions provide rules and guidance for future courts. The custom of using earlier judicial decisions to guide later ones provides stability to the legal system. When deemed appropriate, judges will overturn or modify precedents, but it is generally agreed that this should be done only if there are compelling reasons to do so. It is possible, however, for someone to reject sound reasons for overturning a precedent by claiming that the precedent shouldn't be overturned *only because it is a precedent*. In that case, the one arguing for keeping the precedent would be committing the irrelevant appeal to tradition.

The irrelevant appeal to authority is a fallacy in reasoning in which one argues that a practice or belief is justified because some authoritative person or text asserts it. If a practice or belief is justified there must be good reasons for it and those reasons should explain why the tradition continues or the authoritative person or text supports it. The irrelevant appeal to authority differs from the *appeal to an irrelevant authority*. An example of the latter would be appealing to the advice of an actress with no education or background in medicine to justify seeking some offbeat cancer treatment or for claiming that common vaccines would be harmful to children.

While searching for a fresh example of the irrelevant appeal to authority, I happened upon an arguer who tried to counter an appeal to authority with an appeal to tradition. The arguer was trying to defend fortune telling against those who rejected it because of the authority of the Bible.

> Christianity sees divination as going against the Bible's mandate not to seek "soothsayers," because that would be expressing a lack of faith in God as omnipotent and all-knowing. Yet many...of the world's religions and cultures have woven it into their fiber—Hinduism uses Vedic astrology to match marriage partners; in Chinese culture, an expert is consulted on the most mundane to crucial life matters—from

when to get married to where to live. Wanting to know what will happen is not just a result of our modern brains grasping for control and answers; it's been the human condition for millennia, people have been seeking prophecies since Greeks took often long journeys to consult the Oracle at Delphi (Valerie Reiss, "5 Things to Know Before Going to a 'Psychic'," tinyurl.com/p98vbo).

Reiss argues that since divination has been practiced for millennia in various cultures, it must be good despite what some Christians might say is forbidden by the Bible. The fact that some cultures have been engaging in magical and superstitious thinking for thousands of years does not justify the practice, any more than thousands of years of slavery or genital mutilation would justify those practices. Humans have been beating each other to death in boxing matches for millennia, but that hardly justifies the practice.

The fact that Vedic astrology is still practiced in Hinduism isn't a good reason for thinking that this is a good thing. In fact, it's a bad thing. There is no compelling evidence that any kind of astrology is useful for divining the future, and the belief in this superstition is an open door to fraud, corruption, or worse. The documentary "Guru Busters," produced and directed by Robert Eagle, exposes one of the corrupt astrologers in India who then asked his followers on national television to kill those who exposed his scam.

Reiss doesn't mention what experts the Chinese consult, but it is apparent that she is referring to various kinds of soothsayers. These "experts" bank on the ignorance and superstition of their clients. Perhaps a couple doesn't need any kind of traditional expert to advise them on when to get married or where to live. On the other hand, after the failed assassination attempt on her husband, Nancy Reagan consulted an astrologer to help plan Ronald Reagan's schedule. Donald Regan, one of Ronald Reagan's White House aides, claimed in *For the Record* that "Virtually every major move and decision the Reagans made during my time as White House Chief of Staff was cleared in advance with a woman in San Francisco (Joan Quigley) who drew up horoscopes to make certain that the planets were in a favorable alignment for the enterprise."

Knowing such things makes me proud to be alive in such an enlightened age! We can always hope that Regan exaggerated to help sell copies of his book.

Anyway, I hope that Ms. Reiss is not advising 21st century people to return to the ways of the ancient Greeks. I doubt if too many modern Greeks consult temple oracles for advice on anything, but if they did they might consider that there are much better ways of getting information about the future. A good bit of knowledge has been gained in the past several thousand years and none of it by soothsaying. Using that knowledge to reason inductively about the future, guided by techniques that have been refined over many centuries, has proven to be vastly superior to any form of divination provided by psychics or other fortune tellers. Yet, Reiss is not alone. Surveys in the United States, Australia, and many other "advanced industrial nations," have repeatedly found that most adults believe in astrology or something equally unscientific and magical.

The number of years that something has been practiced, in itself, does not justify that practice. The fact that magical thinking and superstition persist in many areas of modern life does not mean that magical thinking and superstition are superior to other methods. Rather than be guided by the inferior methods of our ancestors, we would be better off if we tried to understand why these primordial ways of evaluating experience persist and what we might do to overcome the tendency to think like our less knowledgeable predecessors. Rather than celebrate ancient errors, we might do better to train ourselves in ways of overcoming our tendencies to fallacious thinking.

One wonders why Ms. Reiss doesn't see that even though the Christians base their aversion to soothsaying on an appeal to authority, their counter-tradition nullifies her appeal to tradition. Or is Ms. Reiss arguing that three traditions trump one tradition? If she is, she's also committing the *ad populum* fallacy. She reminds me of the advertisement that tries to persuade us to buy a particular brand of toothpaste because it's recommended by "four out of five dentists surveyed." What matters is not *how many* dentists recommend anything but *why* they recommend it. Getting paid by a company to recommend it isn't a relevant reason to buy the product. Even having ingredients x, y, and z that promote healthy dental hygiene, while relevant, isn't a sufficient reason for buying one brand rather than another. If another product has the same ingredients, doesn't have anything else that might be harmful, and is less expensive, why not buy the other brand? We might also like to know who did the survey, how did they do it, and a few other things that no advertisement will ever reveal.

It's often the case that arguers combine the irrelevant appeal to authority with the irrelevant appeal to popularity. If it is irrelevant to appeal to one authority to prove a point, then it is irrelevant to appeal to many authorities to prove the same point. You can add zeroes from now until doomsday and you'll still have nothing to show for it. However, it is not always irrelevant to appeal to authorities. If you know nothing about medicine and your physician goes over the results of a medical test with you and recommends a course of action, you are not committing the fallacy of irrelevant appeal to authority when you justify taking that action because your physician recommends it. You might consult another physician for a second opinion, but you would be foolish to consult, say, the janitor or the local newspaper's astrologer. We must rely on experts sometimes, but it is true that experts don't always agree with each other. If, for example, your medical test involved some back problems you've been having, you might get five different opinions from five equally competent physicians on what course of action would be best for you. Why? Recommendations for back problems are notoriously controversial. It would obviously be silly to claim that one recommendation must be the best one since it was made by an expert when there are five different recommendations from five equally competent experts. Ultimately, you should consider all the pros and cons of each of the recommendations and select the option that seems best to you. On the other hand, if four of five equally competent physicians recommend the same course of action, unless you can find a compelling reason for rejecting their position, it would seem that the reasonable course of action would be to follow their advice.

When the majority of experts in a field agree on something, we say there is a consensus. Such is the case with climate experts on the issue of anthropogenic global warming. There are many people, some of them scientists, who do not agree with the consensus that human activities such as deforestation and burning of fossil fuels that result in more greenhouse gases like carbon dioxide are causing changes in our planet's climate that may prove devastating and irreversible. One tactic of the climate change deniers is The Petition Project, which features over 31,000 scientists signing a petition stating "there is no convincing scientific evidence that human release of carbon dioxide will, in the foreseeable future, cause catastrophic heating of the Earth's atmosphere." It is true that 31,000 scientists is a large number, but it is irrelevant to the issue of whether humans are largely responsible for climate change. Most of these 31,000 scientists aren't experts in climate science and, in this case, that matters because when anyone speaks outside his or her own area of expertise their view carries no more weight than that of any other non-expert. What makes it reasonable to accept anthropogenic climate change is not the fact that 95% of all climate scientists agree. It's *why* they agree. Even non-experts can figure out that the experts agree: a survey of all peer-reviewed abstracts on the subject 'global climate change' published between 1993 and 2003 showed that not a single paper rejected the position that global warming is largely caused by human behavior. Climate scientists are not arguing about *whether* global warming is happening. They're not arguing about *whether* humans are largely responsible for global warming. They may be arguing about what action to take. In that case, they should be considered as advisors by those who make policy. Unfortunately, many of those who make policy seem to be ignoring the climate scientists in favor of beliefs pushed by gas, oil, and other corporate interests. Those interests should be considered, but not to the exclusion of the science experts.

A similar tactic has been taken by a group of religious scientists who believe that the scientific fact of evolution is inconsistent with their interpretation of the Bible. The Discovery Institute ran an advertisement with the heading "A Scientific Dissent from Darwinism" in 2001. ("Darwinism" is not a scientific term, but a polemical one used by anti-evolutionists to mean *evolution*.) The ad had over 700 signatures of "doctoral-level scientists and engineers" and stated that "Darwinism," i.e., evolution, isn't sufficient to account for the complexity of life. According to the Discovery Institute, only an intelligent designer, i.e., a god, could account for the complexity of life. Even if that were true, appealing to a group of scientists and engineers who agree with it isn't relevant to the claim's truth.

The *ad populum* tactic has been conjoined with another fallacy by both the climate change deniers and the evolution deniers. Both groups claim that there is a "controversy" over their favored issue. The fact is, however, that you need more evidence than a list of people who disagree with a claim to make that claim "controversial." Otherwise *any* claim could be made controversial just by gathering a few signatures at the mall. The people calling themselves *9/11 Truthers* do not make what happened to the twin towers controversial by forming a group called "Architects and Engineers for 9/11 Truth." (The ease of making people think something is controversial was demonstrated by Penn and Teller in an episode of *Bullshit!* in which they had some people gather signatures to ban the dangerous chemical *di-hydrogen oxide.* Many people die each year from this chemical. Told of its dangers, many people eagerly signed the petition to ban H_2O or water.) Global warming and evolution would be controversial if the majority of scientists in those respective fields—climate science and biology—were arguing about whether there is global warming or whether evolution of species had occurred. This is obviously not the case, nor is it the case with the Apollo Moon landing, the safety of vaccines, HIV as the cause of AIDS, and the danger of cigarette smoking. These items do not become controversial just because you can find one person or many persons who disagree with the consensus. The fact that there is a consensus means that the issue is *not* controversial. The careful reader understands that being non-controversial is not the same as being true. If you don't understand that, then you did not understand what I wrote above about how what matters is not *how many* experts agree on something but *why* they agree.

It's not always easy to determine whether a premise is relevant to proving a claim. What makes a premise relevant or irrelevant depends on the argument. No claim is relevant or irrelevant in itself. Relevance of premises is always relative to proving or supporting the conclusion. Whether or not a particular claim is relevant to a conclusion depends on the subject matter and on exactly what the argument is about.

Relevance should not be confused with *significance*. Two pieces of evidence may both be relevant to a particular position but one may be more significant than the other. For example, it may be true that a killer wore a black hat and an expensive set of gloves and that he left a unique and peculiar shaped footprint at the scene of the crime. The fact that the suspect owns a black hat is relevant but relatively insignificant, since many people own black hats. But the fact that the suspect owns a pair of expensive gloves matching those of the killer and his footprint matches exactly the unique and peculiar footprint taken from the scene of the crime are not only relevant but very significant.

It's relevant to bring up the fact that shortly after a vaccination your child was diagnosed with autism. Of course, having your child diagnosed with autism is significant to you. But in terms of evidence, the many scientific studies that have found no causal connection between vaccinations and autism carry more weight than a single personal experience of one thing happening after another.

There is, however, very little specific advice to give regarding the evaluation of the relevance of premises. The best general advice is to *avoid the common fallacies of relevance*. We've already looked at a few of these common fallacies. Here are a few more.

Irrelevant comparisons. Many advertisements make irrelevant comparisons, e.g., comparing a relatively inexpensive single-function Minolta copier to an expensive, multi-function Xerox machine. True, both make good copies—and the ad tries to get the consumer to focus on this fact—but the Xerox machine performs a multitude of tasks (such as collating, stapling, double-sided copying, etc.) which the Minolta cannot do. These differences are ignored in the ad. Thus, even though the ad is correct in stating that copies made on the Minolta will be about equal in quality to those on the Xerox, but much cheaper per copy, the comparison is irrelevant. To be relevant, the Minolta should be compared to an *equivalent* Xerox machine.

Another ad that uses an irrelevant comparison is the one that compares two brands of paper towels by dropping an egg and trying to catch it with the paper towel. Since paper towels generally are not used to catch eggs, the fact that one is better than the other at this task is irrelevant. More important, the texture of such a towel might be such that it is less absorbent than other paper towels. Since paper towels are used to wipe things up, it might be the case that what makes the towel capable of catching eggs also makes it less capable of absorbing spills.

Articles on the high cost of going to college also often commit the fallacy of *irrelevant comparison.* It is irrelevant, for example, to compare the cost of going to a public community college with the cost of going to a private university. Room and board costs are not included in the community college costs, but they are in the private school costs. If one is going to compare costs, then it is not relevant to consider room and board costs for one type of school but not the other. Furthermore, the institutions are extremely different. A more relevant comparison would be to compare the costs of various private universities to one another or the costs of various community colleges with one another. Anyway, projections about future costs are risky. The forecaster can't be sure that relevant economic conditions, such as the rate of inflation, will be the same in 18 years when a newborn might be starting college. Telling parents they'll need millions of dollars to send their new kid to college, based on a high rate of inflation at the time of writing, should come with a warning: *my speculation is true if the inflation rate continues as it is for another eighteen years.*

Irrelevant appeals to feelings or emotions. Arguers often make *irrelevant* emotive appeals when they lack logical reasons for their position or when persuasion rather than truth is their goal. Emotion is generally a much more powerful *motive* to belief and action than logical reasons are. Hence, emotive appeals are often persuasive. Some people are moved to purchase products they do not need by irrelevant appeals to fear, guilt, vanity, or the desire for pleasure. Despite what the advertisers repeatedly tell us, our children are not going to grow up to be imbeciles if they don't have the latest computer, we are not bad parents if we don't supply our children with brand-name clothes, and they will not grow up without any friends if they don't wear those brand-name items. Furthermore, your hair will not suddenly become thick and beautiful when you switch shampoos. Nor will people suddenly find you attractive when you buy a new car or switch aftershave lotions.

Others are vulnerable to appeals to pity. Show them a tear and they open their pocketbooks. It is difficult to overcome our natural instinct to follow our emotions and reason ourselves away from the pull of strong emotive appeals. But just as nobody in advertising ever went broke bypassing the intellect and aiming straight for the pleasure principle, nobody ever said it would be easy to be a critical thinker. I know it's hard to believe but blondes don't necessarily have more fun and that beautiful woman standing next to the car in the ad doesn't come with the purchase. Nor do lovely ladies fall out of windows when you pass by with your special cologne on.

Of all the emotions that fuel the fallacy-driven life, none can compare with fear. The anti-vaccination scare discussed in the previous chapter was fueled by mass media reports that can only be described as scandalous and incompetent. Millions of parents who should know better are not having their children vaccinated out of fear that vaccines might cause autism in their children. But the anti-vaccination movement didn't begin out of fear. It began when some media-savvy parents made a hasty conclusion based on incomplete evidence that found a causal connection unsupported by the science. The movement got traction when Andrew Wakefield's work was given much more media attention than it deserved. Actually, the media attention was appropriate, but it was misdirected. Instead of criticizing Wakefield for making claims that were unsubstantiated, the media gave him a bully pulpit to promote his claims. He must be a likeable enough fellow, though, because he still has many parent advocates in the United Kingdom and the United States, despite having had his license to practice medicine revoked. And if it wasn't bad enough to have a movement that encouraged parents not to have their children vaccinated against diseases that can maim and kill, the leaders of the anti-vaccination movement took it a step further and advocated distrusting scientific research itself. As the evidence accumulated that there is no causal connection between vaccines and autism, the parents and their medical leaders who started the movement dug in deeper. Rather than accept what the science was revealing, the leaders of the anti-vaccination movement began questioning the motives of the scientists. The scientists are in it for the money, they charged. Of course scientists don't find proof that vaccines cause autism; they're in the pay of Big Pharma. Some—like actress Jenny McCarthy, who became a leader of the movement after claiming she had cured her son of autism—claimed that "mother warriors" like her knew the truth, while scientists were intentionally producing faulty work to show there's no connection between vaccines and autism. (*Mother Warriors* is the title of one of McCarthy's books.) So, while it may be fear that is motivating most parents who refuse to have their children vaccinated, it was ignorance and questionable assumptions (Jenny McCarthy), greed (Wakefield), faulty causal reasoning (Lyn Redwood), and *ad hominem* attacks (all of the above) on the motives of those who produced evidence contrary to their beliefs, that initiated the anti-vaccine movement.

I'll finish this section on fallacies of irrelevance with some examples of poor, but common, arguments that all fall under the heading of the *non sequitur*. Each premise in the following arguments is irrelevant to its conclusion and none of the conclusions in these arguments follows from the premises.

"I'm going to a homeopath because my medical doctor gave me the wrong prescription."

"We've been unable to crack the case, so we hired a psychic to help us with our investigation."

"We hired the psychic for our investigation because we have to leave no stone unturned."

"I'm going to an acupuncturist for my migraines because the homeopath didn't help."

"I can't figure out how any natural process could account for the complexity of the human cell, the bacterial flagellum, or a donkey's eye; therefore, an intelligent designer must have put these parts together in the cell, the flagellum, and the donkey's eye."

Assumptions, Omissions, and Illusions

Appealing to irrelevant reasons isn't the only kind of fallacy we humans engage in, of course. We often make false or questionable assumptions. We often draw conclusions based on insufficient evidence. One of our most favored fallacious moves, however, is to omit all the evidence that would count against our position. We tend to be selective in what we present to others to get them on our side. Selective thinking is our natural way of thinking. We are drawn to evidence that supports what we already believe. We are not very attracted to evidence that goes against our beliefs or our emotional needs. Psychologists call this natural tendency to be selective in both our memory and our perception *confirmation bias*. People with strong convictions often take confirmation bias to a level known as *motivated reasoning*. The more evidence one presents against their belief, the more motivated they become to refute the evidence and defend their conviction. I have mentioned several examples of motivated reasoning, e.g., that vaccines cause autism, that evolution didn't happen, that climate change isn't happening, and that 9/11 was an inside job of the Bush administration. Other examples include the belief that Barack Obama was not born in Hawaii, that AIDS is not caused by HIV, that the Apollo Moon landing was a hoax, and that the Holocaust never happened.

It is a good lesson in critical thinking to observe how easily intelligent people can see intricate connections and patterns that support their viewpoint and how easily they can see the faults in viewpoints contrary to their own. To try to falsify a belief we are certain of goes against our nature. As long as one ignores certain facts and accepts speculation as fact, one can prove just about anything to one's own satisfaction. It is a difficult but necessary requirement of good thinking to try to *falsify* a pet belief. If you believe that a full Moon will, say, cause more people to be admitted to the emergency room at your hospital, you will pay particular attention if the hospital is busy on the night of a full moon. But on other nights when the hospital is just as busy, you won't attribute the activity to the moon. (On 19 November 2011, there were nine people shot in four separate incidents in Sacramento, California. The Moon was not full, but nobody seems to have noticed.) A critical thinker does the unnatural thing and collects data to check her intuition against the facts. Some scientists have actually done this and guess what? They haven't found a significant correlation between phases of the Moon and hospital admissions in the emergency room. Nor have researchers found a significant correlation between any phase of the Moon and any of the following: the homicide rate, traffic accidents, crisis calls to police or fire stations, domestic violence, births of babies, suicides, major disasters, casino payout rates, assassinations, kidnappings, aggression by professional hockey players, or violence in prisons. That the facts go against a common belief is not so exceptional. But that people are reluctant to accept facts presented to them indicates just how unnatural critical thinking is. The critical thinker has to be willing to give up pet beliefs when the data do not support the beliefs. As noted above in discussing the backfire effect and the continuing influence effect, changing one's mind even when presented with strong evidence that one is clinging to a false belief does not come naturally to many people. Many of us are guilty of a variant of what is called the *fundamental attribution error*: we attribute our own and the beliefs and actions of those who agree with us to high-minded, objective, unbiased inquiry, while we attribute the beliefs and actions of those who disagree with us to bias and ulterior motives. The *illusion of principled inquiry* is only one of many illusions that hinder critical thinking.

Another common illusion is *the clustering illusion*, discussed in Chapter Seven where the focus was on falsely identifying clusters of cancers as evidence of an environmental cause. A classic study done on this illusion demonstrates just how hardheaded we are when it comes to facing facts that don't support our beliefs. The study was done by Thomas Gilovich and some colleagues. It centered on the belief in the "hot hand" in basketball. It is commonly believed by basketball players, coaches, and fans that players have "hot streaks" and "cold streaks." A detailed analysis was done of the Philadelphia 76ers during the 1980-81 season. It failed to show that players hit or miss shots in clusters at anything other than what would be expected by chance. Gilovich et al. also analyzed free throws by the Boston Celtics over two seasons and found that when a player made his first shot, he made the second shot 75% of the time and when he missed the first shot he made the second shot 75% of the time. Basketball players do shoot in streaks, but within the bounds of chance. It is an illusion that players are 'hot' or 'cold'. When presented with this evidence, believers in the "hot hand" are likely to reject it because they "know better" from experience.

We humans don't really like to face facts. Our natural reaction when you tell us something we don't want to hear is that you're wrong. Worse, many of us spend most of our lives defending what is most likely not true. Worse still is that we don't seem to care that we live our lives defending the improbable while wearing impenetrable belief armor that protects us against all challenges. No matter what nonsense we believe, others have their own nonsense that they are sure is better than ours, and you will not be able to persuade them otherwise. Is this a good thing? Is there any hope for us to overcome these natural tendencies to bias and fallacious thinking? Yes, there are at least 59 ways to overcome our biases and errors, as you will see if you stick around and read the penultimate and final chapters of this book.

SOURCES FOR CHAPTER EIGHT   275

IX
Are We Doomed to Die with Our Biases On?

"Societies progress by the free assertion of differing proposals, followed by criticism, followed by the genuine possibility of change in the light of criticism....The whole approach of an authoritarian society is anti-rational. A rational and scientific approach requires societies to be open and pluralistic."—Karl Popper

In this, our penultimate chapter, we review and explore why we believe so many things that are palpably not true.

The Joy of Jumping to Conclusions

All of us believe things that are palpably not true and most of us believe many other things that are probably not true. We naturally and instinctively gravitate toward beliefs that make us comfortable or fit in with the beliefs of our families and friends. It's understandable that we would believe many things as children that our parents and others in our society have passed on to us as if they were absolute truths, even though they may be nothing but traditional prejudices. But why do we cling to what Mencken called the "palpably false" after we're old enough to think for ourselves? Have we been brainwashed by our parents, teachers, clerics, and friends? Those of us who live in so-called advanced industrial societies might blame the inundation of repeated messages in the media, advertising, television, films, talk radio, or the Internet. But there is precious little evidence that our brothers and sisters in the so-called Third World have been spared belief in untruths along with the message blitzing to which we are subjected from cradle to grave. Is it in our genes, our cultural upbringing? Is there something in the way we go about evaluating our experience and accumulating beliefs that leads us astray?

Consider the following facts. We have scientists who are taken seriously when they proclaim that a parrot and a dog are psychic. More than half of all Americans believe we can heal each other by psychic or spiritual means. About one-third believe in telepathy and about one-fourth believe in clairvoyance. More than one-third believe houses can be haunted. More than forty percent accept demonic possession as real. Twenty to twenty-five percent believe that the dead communicate with us. Only about forty percent of Americans accept evolution and about fifty percent believe a god created humans in their present form. Only one-third of Americans accept the Big Bang theory. Virtually all scientists accept both evolution and the Big Bang as facts. The average citizen is not impressed.

About one-third of Americans and Australians believe UFOs are spacecraft from other planets (see Appendix B for a consideration of some of the reasons for not thinking aliens have visited us). Many people believe cell phones cause brain cancer, despite a preponderance of evidence to the contrary (see Appendix A). Likewise, many people believe vaccines cause autism, despite a lack of scientific evidence in support of that belief (see Chapter Seven). And, despite easily accessible and overwhelming evidence to the contrary, a majority of American adults still believe Iraq had weapons of mass destruction before the U.S. invaded that country at the order of George W. Bush (see Chapter Six).

There is some comfort in knowing that most people have come to their beliefs about scientific matters without knowing or studying the science. But there is little comfort in the fact that many people don't seem to care that they are ignorant or unqualified to make judgments about such matters. We have a "penchant for jumping to congenial conclusions," noted psychologist Barry Beyerstein. We don't seem to be aware of the fact that our natural faith in memory and perception should be tempered by a fear of error and a concern for accuracy. Care about the truth should motivate us to be more careful about blindly accepting what seems obviously true based on personal experience or what we learned at our mother's knee. On the other hand, it is perfectly understandable why a parent whose child showed symptoms of autism shortly after getting a vaccination would make a connection between the two. There is little mystery as to why a young man might think he has precognition when a healthy aunt drops dead a day or two after the young man dreamed something terrible happened to his aunt. It is not that hard to understand why most of us believe strongly in things that are palpably not true. We have a natural propensity to see causal connections where there are none and our beliefs are constantly reinforced by people we like and trust. For many people, one vivid and emotionally salient experience validated by a single neighbor or shopkeeper trumps a thousand randomized, double-blind, controlled scientific experiments.

I am familiar with most of the arguments in defense of astrology and against evolution, in defense of UFOs as alien spacecraft and against the view that humans are largely responsible for global warming, in defense of various alternative therapies (acupuncture, aromatherapy, chiropractic, detoxification, homeopathy, naturopathy, and various forms of energy medicine and unscientific cancer cures) and against science-based medicine, in defense of the view that vaccines cause autism and against the view that cell phones are safe. Since 1994 I have operated a website called The Skeptic's Dictionary (www.skepdic.com) where I have engaged in arguments on all these issues in a public forum. Over the years I have gradually become less interested in persuading others and more interested in understanding why people believe the things they do. Many skeptics attribute belief in things that are palpably untrue to psychological mechanisms such as *cognitive dissonance* or to personality disorders such as *the fantasy-prone personality*. I don't find either of these explanations useful or supported by compelling reasons. (For my reasons see <skepdic.com/cognitivedissonance.html> and <skepdic.com/fantasyprone.html>.)

Many believe palpably untrue things because they were brought up to believe them and the beliefs have been reinforced by people they admire and respect. Many believe palpably untrue things because they are ignorant of such things as fundamental biological or physical processes. They don't know much about blood and nutrition, for example, so the blood-type diet—that one's diet should be determined by one's blood type—seems plausible and convincing to them. Many believe certain things because they follow faulty rules of reasoning. For example, they think personal experience should trump scientific studies or they judge the beliefs of those who disagree with them as being motivated by flawed personal characteristics.

Many of the beliefs I have been calling *palpably untrue* are called *weird things* by Michael Shermer, the author of *Why People Believe Weird Things*. Shermer claims that "More than any other, the reason people believe weird things is because they want to....It feels good. It is comforting. It is consoling." Secondly, weird beliefs offer 'immediate gratification." People like weird beliefs because they are simple. Weird beliefs also satisfy the quest for significance: they satisfy our moral needs and our desire that life be meaningful. Finally, he says, people believe weird things because weird things give them hope. I think Shermer is spot on for certain kinds of beliefs. For example, it seems apparent that many people believe that "mediums"—like John Edward, Sylvia Browne, and James Van Praagh—get messages from the dead because *they want to believe so they can connect to a deceased loved one and get validation that there is life after death.* In any case, what's consoling to one person is distressing to another, e.g. the belief that children born with cancer are part of a divine plan that makes sure justice is always done in the end and that everything happens for a reason.

On the other hand, communal reinforcement, ignorance, and wishful thinking don't seem to explain why someone like Deepak Chopra, a trained medical doctor, would believe that thoughts can affect biological processes at the molecular level. "Happy thoughts make happy molecules" is one of the ways Chopra expresses this belief. There is no compelling scientific evidence for this belief. I can only conclude that educated, knowledgeable people who believe such things are following a faulty rule of reasoning or speaking metaphorically. (If all Chopra means is that mood affects stress and reduction of stress releases hormones that can be beneficial, then he is saying nothing new and nothing profound.) In fact, I have found that many of the people I've argued with over beliefs that I consider palpably untrue are not irrational, but believe what they do for the same reason I believe what I do: *they think the evidence supports their position.* Sometimes my opponents and I differ about how to treat a particular kind of evidence, e.g., a medical doctor who defends homeopathy did battle with me over the evidence for homeopathy and he had quite a different understanding than I do of what it means for something to be "statistically significant." Many times, however, I have found that my opponents and I have differed about what rule to follow when determining what *weight* to give different kinds of evidence. One correspondent put it this way: "My son was diagnosed with PDD-NOS (Pervasive Developmental Disorder—Not Otherwise Specified) a year ago, and he is the only peer-reviewed double-blind placebo controlled study that I need." A similar claim was made by the chiropractors described in Chapter Two who would not give up their belief in applied kinesiology after it failed a randomized, double-blind trial. (That particular test wasn't actually a randomized clinical trial, but for the sake of illustration here we'll treat it as one.) Their response was to assert that randomized controlled trials (RCTs) don't work. How did they know that? The results of the RCT didn't jibe with what they believed they knew to be true from experience. Actress Jenny McCarthy typifies this way of thinking. After declaring that she had cured her son Evan of autism—which she blamed on vaccines—she announced that Evan was her science. I have found that one reason my opponents and I could rarely come to an agreement was that many of them give more weight to personal experience than to scientific studies. They have more trust in anecdotes and in their interpretations of those anecdotes than in RCTs, unless those RCTs support their position. Should they? I don't see why they should. There are very good reasons why we should be more trusting of studies done by experts without any personal interest in their outcome than we should be of interpretations by non-experts with a strong personal or emotional stake in the outcome. I went over these reasons in Chapter Seven and won't repeat them. Suffice it to say that the self is not the most unbiased observer of its experiences and to be fair one must consider *all* the relevant evidence.

Balancing the Data of Experience and the Data of Science

Science is not infallible, but it is more likely to do a better job of identifying patterns and determining causal relationships than the untutored layperson evaluating a highly charged emotional issue like the origin and nature of a disease that has afflicted one's child. I admit that there have been times when the scientists got it wrong and the parents got it right regarding both cause of a disease and the proper treatment of the afflicted. Leo Kanner and Bruno Bettleheim convinced many psychiatrists and laypeople to adhere to the discredited notion that lack of maternal warmth causes autism and schizophrenia. There are many in medicine who have much to be ashamed of for advising parents to institutionalize any baby with Down syndrome. And the Temple Grandin story reminds us that at one time the experts recommended that those on (what is now called) *the autism spectrum* be institutionalized. When Temple Grandin (b. 1947) was a young child she didn't speak, she didn't like to be touched, and she wouldn't make eye contact. A doctor told her mother that Temple should be institutionalized. Her mother kept her out of institutions, hired a speech therapist and a nanny, sent her to various special schools, got her through high school, and sent her off to college. Grandin went to graduate school, became interested in caring for cattle and humanely preparing them for the slaughterhouse, began writing articles about cattle facilities, and eventually earned a Ph.D. in animal science. Today, one-third of the cattle and hogs raised in the United States are handled in facilities Temple Grandin designed.

The fact that science and scientists are fallible shouldn't be taken as a sufficient reason to distrust science in general or for trusting your own instincts above those of science. There have been an overwhelming number of high quality scientific studies regarding a possible causal connection between vaccines and autism, and between cell phones and brain cancer. To trust one's instincts or to appeal to a distorted version of the *precautionary principle*—the idea that if there is some chance that someone somewhere at some time might be harmed in some way, then we should not go ahead with a practice—when there is overwhelming evidence that a practice isn't harmful, does not seem rational. You could justify banning just about anything on the grounds that somebody somewhere might be "especially sensitive" to it and be harmed by it. You could ban driving, cooking, heating anything, and a host of other wonderful activities if preventing anyone from being harmed under any circumstances were our guiding principle.

One of the first things I learned when I began teaching courses in critical thinking in 1974 was that those of us trained in philosophy needed to supplement our philosophical training with the study of various cognitive, perceptual, and affective biases or illusions. We had a lot to learn from the social scientists if we were to teach our students to think critically. Identifying fallacies, learning to test deductive arguments for validity, and the like would not be enough. Mastering those tasks would not be enough, that is, unless our goal was to send out into the world a generation of smart-sounding but inadequately trained poseurs. Students should be taught how the mind works and why we humans—even the smartest and most educated among us—continue to make the same kinds of reasoning errors generation after generation despite all our knowledge and training. Without awareness of common pitfalls such as confirmation bias, positive-outcome bias, and subjective validation, a person trained in logic and fallacy detection is easily deceived into thinking that he or she has acquired invincible armor against assaults of unreason. Expressions like *post hoc ergo propter hoc* and *false cause*, should be informed by knowledge of evolution and how the brain works to jump to conclusions about causal connections. That knowledge should be a basis for explaining *why* science uses methods like the randomized, double-blind experiment in the search for causal relationships, and *why* science uses the methods it does to investigate alternative causal mechanisms.

Our Biased Brains

Thirty-five years ago, when I began using Howard Kahane's *Logic and Contemporary Rhetoric* as a textbook in my logic classes, there weren't many books written for the general public that focused on the various ways we are deceived and misled by assumptions we are prone to make about thinking, perception, and the testimony of others. In recent years, several excellent books have filled this gap. (See the sources listed for this chapter for the titles of several such books.) *The Invisible Gorilla: and Other*

Ways Our Intuitions Deceive Us is one such entertaining and educational book. The authors, Christopher Chabris and Daniel Simons, explain how the mind works. Guess what? The mind doesn't work the way intuition tells us it works. Intuition tells us that the mind presents the world as it is. Actually, the mind constructs the world we perceive. The mind doesn't construct the world out of nothing and the world is not just an illusion. But neither is the mind working like a video recorder. Much of what the mind presents to us, though, is an illusion. The mind is constantly tricking and deceiving us—for our own good—both perceptually and cognitively. After several chapters of mostly bad news about the trials of overcoming perceptual and cognitive illusions, the penultimate chapter of *The Invisible Gorilla* brings some good news: there are some exercises you can do to strengthen your brainpower and thereby reduce your chances of being deceived by the same brain you are trying to train.

If you were thinking of Brain Gym, think again. It won't help, nor will knowing your brain type. Neither will listening to Mozart or taking Procera AVH. Nor will using subliminal tapes or hypnosis. Your self-help guru won't provide you with much more than the *illusion of control*. The scientific evidence shows that about the only thing that consistently leads to improved brain function is physical exercise. Of course, if you're leading an unhealthy life most of the time, even daily workouts at the gym won't help you much.

The bad news is that most of the folks claiming to have the key to unleashing your inner brain so that you can reach your mythical "true potential" are trying to sell you an illusion. Chabris and Simons call it the *illusion of potential*. Steve Salerno wrote a whole book about the sellers of this illusion, the movers and shakers in the *human potential movement*. He called it *Sham*. The stars of this movement include Richard Bandler, Werner Erhard, John Grinder, L. Ron Hubbard, and Tony Robbins. The programs go by many names. A few of the more popular are: est, Landmark Forum, neuro-linguistic programming, Tony Robbins seminars, Impact Training, MJB Seminars, Silva Mind Control, the Demartini Method, The Work (Bryon Katie), PSI Seminars, Mind Dynamics (the daddy of them all), Lifespring, Hoffman Quadrinity, and

Complete Centering. (See Appendix E: How to Create Your Own Pseudoscience.)

Chabris and Simons review several examples of how to "transform a claim with almost no scientific support into a popular legend that fuels multimillion-dollar businesses." They go into great detail explaining the origin and success of the notion that listening to the music of Wolfgang Amadeus Mozart will increase your child's IQ (the so-called *Mozart effect*). They're especially qualified to do so: Chabris published a meta-analysis of studies that allegedly support the idea that listening to Mozart increases one's IQ. Chabris concluded that rather than proving that listening to Mozart was beneficial, the studies could just as well be used to argue that sitting in silence or relaxing makes you dumber. The authors found similar exaggeration behind Nintendo's overly hyped Brain Age software for gaming systems.

About the only uncontested effect of cognitive training is that training in a specific area improves performance in that area but does not transfer to other cognitive tasks. Even learning to memorize long lists of *numbers* doesn't help one learn to memorize long lists of *letters*. "Practice improves specific skills, not general abilities." So, put down that Sudoku, unless your goal is to get better at doing Sudoku or you just like doing it. If you're trying to exercise your brain, you'd do better to take a brisk walk.

There is some positive news for those who use groups to make decisions: you're better off if you have the members of the group think about the issue independently and bring their written thoughts to the table. Many groups use the "brainstorming" technique that brings the members to the table to have them listen to and discuss each other's thoughts on some problem or issue. The members are told not to be critical of their own or other's ideas until all the ideas have been put on the table. Some groups brainstorm on the fly. This method has been shown not to work as well as having the members come to the table with their independently derived thoughts. The reader can use his or her intuition to figure out why this is the case, or you can read the book for details. (This group dynamic is also discussed in Daniel Kahneman's *Thinking, Fast and Slow*.)

Chabris and Simons are most well known for their work on *inattentional blindness*. Their book begins with a discussion of the so-called "invisible gorilla test," which isn't really a test at all. It's not even accurate to refer to "the invisible gorilla," since about half of those who have been tested see the gorilla in the film they are asked to watch. (As noted in Chapter Three, the viewers are asked to pay attention to something that is going on in the film: the passing of a basketball among several players in white shirts being guarded by several players in black shirts. A person in a gorilla suit walks through the scene, faces the camera, and walks off. The interesting thing about the test is that about half of those who take it don't see the gorilla. Many adamantly maintain that *two* films were used and that they were tested with the film that had no gorilla. Most who don't see the gorilla the first time through are shocked that they could have missed something so salient crossing their visual field without noticing it.) Chabris and Simons tell us that despite what some people think, there is no difference in personality, intelligence, or educational level that distinguishes those who see the gorilla from those who don't. There's also no gender or age difference between those who see it and those who don't. (When I showed the film to my classes, I tried the trick of telling them that females usually do better than males, just to give both groups an incentive to focus more intently on the task at hand. It made no difference.)

The fact that no significant difference in intelligence, education, or personality has been found in those who do and don't see the gorilla right before their eyes might be of some comfort. On a similar note, there is little evidence that intelligence or education can account for why people believe differently about topics such as vaccinations, global warming, cell phones and cancer, paranormal phenomena, UFOs, alternative and science-based medicine, and the like. When I refer to something like homeopathy as 'absurd' I am not intending to imply that homeopaths or their patients are stupid. Their beliefs have little to do with intelligence and my criticisms of their beliefs should not be taken as making judgments about the intelligence or integrity of the holders of those beliefs. I have no doubt that many of the people I've argued with publicly about the topics just mentioned are much smarter than I am. In fact, there is good scientific evidence to support the notion that being really intelligent and knowledgeable can be a disadvantage to some thinkers because of the increased ability to come up with rationalizations in defense of a position one originally adopted for inadequate reasons. There are many reasons why smart people sometimes believe dumb things. The smarter one is, the easier it is to see patterns, fit data to a hypothesis, and draw inferences. The smarter one is, the easier it is to explain away strong evidence contrary to one's beliefs. Also, smart people are often arrogant and incorrectly think that they cannot be deceived by others, the data, or themselves. Many really smart people believe some really dumb things. Medical doctor, cancer specialist, and blogger (Science-Based Medicine and Respectful Insolence) David Gorski wrote of "the Nobel disease," which he defined as "a tendency among Nobel Prize recipients in science to become enamored of strange ideas or even outright pseudoscience in their later years." I investigated the idea and found an even dozen Nobel Prize winners who have advocated ideas that are absurd...excuse me, are not supported by the scientific evidence. (See http://www.skepdic.com/nobeldisease.html.) I also expanded Gorski's idea and found *three* dozen medical doctors who advocate ideas that are not supported by the scientific evidence. Since there are many more M.D.s than there are Nobel laureates, finding M.D.s defending weird ideas was rather easy. Does that mean there is an M.D. disease? No, but it does show that education and intelligence should not be equated with "doesn't believe nonsense."

As noted above, the so-called gorilla test exemplifies *inattentional blindness*, one of several everyday illusions. Such illusions can have significant effects that run the gamut from mistaken identity to false imprisonment to death on the highway, the runway, or the high seas. Some of these everyday illusions lead to misguided legislation that aims at preventing harm but in fact doesn't prevent harm and may actually lead to *more* harm. Several legislatures, for example, have passed hand-held phone laws for drivers, banning driving while holding a cell phone but allowing driving while talking on a hands-free phone. The legislation is based on the illusion that holding the phone makes it harder to steer a vehicle and thus more dangerous than driving while using a hands-free device. Seventy-seven percent of Americans think it's safer to talk on a hands-free phone than on a handheld phone. The empirical evidence shows otherwise. The evidence shows that the deficit in driving skill has nothing to do with holding or not holding the phone but with the distraction that comes from talking on the phone while driving. The problem is with the eyes, not the hands. The dangers of hands-free phone use while driving might be amplified by another illusion, the *illusion of confidence*, by deluding a driver into thinking that she can drive safely while talking on the phone as long as her hands are free. Despite your belief in your abilities to multitask, "the more attention-demanding things your brain does, the worse it does each one."

Unfortunately, the evidence shows that training people to be more attentive by trying to develop an ability to notice the unexpected doesn't help. We've evolved to notice what we need to notice to survive and multiply. When you read about the young girl killed in a crosswalk while talking on her cell phone or listening to her iPod, don't wonder what happened to her instincts. Her instincts were probably just fine, but our instincts didn't evolve for life with the kinds of distractions we face in the 21st century. Chabris and Simons remind us: "Our neurological circuits for vision and attention are built for pedestrian speeds, not for driving [or flying!] speeds."

The fact that inattentional blindness is unavoidable doesn't mean there aren't important lessons to be learned from studying it. We might be more understanding of people who claim they didn't see something even though it was right before their eyes. They may be telling the truth. On the other hand, we might be less taken aback when we notice something that is right before our eyes that we didn't notice a few seconds or minutes or days ago. What one takes to be a miracle might just be a matter of inattentiveness while perceiving. Inattentional blindness may explain, for example, how a pilot with an interest in crop circles could fly right over one without even noticing it. The pilot had flown to see a recently discovered crop circle near Stonehenge. After visiting the site, he flew back to the airport to refuel before setting off on a trip that took him back over the site he had just visited. On the return flight he noticed another crop circle near the one he had visited earlier in the day and swears that the new circle was not there just forty-five minutes earlier. The new circle is very elaborate and could not have been produced by human hoaxers in such a short time. He concluded that some mysterious force must have been at work. Perhaps, but it seems more likely that the pilot experienced inattentional blindness when he was flying to the airport. He was focused on other tasks when he flew over the site and didn't notice what was right beneath him all the time.

Awareness of inattentional blindness should also make us more realistic about what we should expect from those manning baggage scanners at airports and from our radiologists or dentists reading x-rays. I don't think we should speculate, however, that inattentional blindness explains why others don't see things the way we see them, as Dean Radin does in his attempt to explain why skeptics reject his and others' work on paranormal phenomena (*Entangled Minds*, p. 44). To assume that those who disagree with you aren't paying attention to the right stuff seems like the height of self-serving arrogance.

Perceptual illusions like inattentional blindness (or inattentional deafness) and *change blindness* (discussed in Chapter Three) reveal some important facts about perception. Our brains have evolved to produce useful representations without requiring faithful duplication of the visual (or auditory, etc.) field available to us at any given moment. The brain isn't storing hundreds or thousands of little details at each moment and constantly comparing those details to see if anything's been missed or has changed. Perception is determined in part by expectation, which makes perceiving the unexpected difficult unless it stands out vividly against the general picture perception provides.

Just as vision does not function like a video camera, memory does not function by recalling replications of faithful representations that have been experienced. Both perception and memory are constructive activities and both are prone to error in the act of constructing a vivid perception or memory. Most of us are deceived into thinking that the more vivid and detailed our memories are, the more accurate they are. The scientific evidence does not support our intuition here. There are many vivid examples of people having vivid but false memories of having been kidnapped, lost in the mall, or meeting some famous person. Studies on memories of salient events like the assassination of President John Fitzgerald Kennedy or the 9/11 terrorist attacks show that many of these memories are inaccurate, despite their vividness and the confidence we have in them. This illusion of confidence not only leads us to put more faith in our own memories than we should, it also leads us to put more faith in the testimony of others when they express certainty and appear self-assured. One of the more difficult points I had in getting my critical thinking and philosophy of law students to accept was the fact that eyewitness testimony has the strongest influence on jurors but is known to be unreliable. Eyewitness testimony is certainly less reliable than physical or circumstantial evidence—such as DNA evidence—even if our intuition tells us otherwise. (The factoid presented in Chapter Three illustrates this point: eyewitness identifications account for about three-fourths of wrongful convictions later overturned by DNA evidence.)

Ignorance and the Illusion of Knowledge

Just as lack of knowledge about perception and memory can hinder critical thinking, ignorance in any specific field will hinder the ability to think critically about matters in that field. You can't think very well about any subject, no matter how intelligent you are, if you don't have the necessary knowledge. On the other hand, many people with a great deal of knowledge prove to be ignorant of what they don't know, and the consequences of that ignorance can be disastrous. It should go without saying that information is not the same as knowledge, otherwise we'd all be extremely knowledgeable. Most of us are overwhelmed with information, but we shouldn't assume that being familiar with dozens of factoids is equivalent to understanding anything.

Having knowledge and thinking one has knowledge are often worlds apart. Whenever people think they know more than they do, they're under the influence of the *illusion of knowledge*. I won't bore the reader with the obvious examples of arrogant but self-confident political leaders who have led us and kept us in wars where there isn't even a sensible meaning to the words 'win' and 'victory.' Most readers are familiar with the cold fusion claims of Pons and Fleishman, Jacques Benveniste's water memory claims in defense of homeopathy, and Prosper-René Blondlot's "discovery" of non-existent N-rays. On the other hand, there is an obvious benefit to the illusion of knowledge. True, it gives us more confidence than we should have and creates an unrealistic view of our world, but it provides us with an optimistic viewpoint that allows us to get out of bed in the morning and go through the day oblivious to what inveterate ignoramuses we actually are. Depressed people, many of you will be happy to know, don't usually suffer from the illusion of knowledge. Still, it is depressing to realize that many people running wars or political campaigns—or betting on horses, commodities, or stocks—are prone to mistake their good luck for skill and knowledge. This has the added negative effect of giving them the *illusion of confidence*. It is a hard lesson to accept, but the success of many people is due to luck, not knowledge. If a thousand people try a thousand different methods and one of them hits the jackpot, it is an illusion to think the winner necessarily had more knowledge or skill than the losers. If two psychics pick opposite winners in an athletic contest, one of them may appear to have more knowledge that the other, but the appearance is an illusion.

Playing on the illusion of knowledge, hucksters use "technobabble" to try to sell us such things as expensive audio equipment or HDMI cables when the evidence shows that there's no meaningful difference between the expensive and the inexpensive stuff. Chabris and Simons are particularly offended by the use of "colorful images of blobs of activity on brain scans that can seduce us into thinking we have learned more about the brain (and the mind) than we really have." They call such appeals "brain porn."

Some might be surprised that the authors of *The Invisible Gorilla* have many good things to say about those weather forecasters who appear on the nightly news to give us their forecasts. We may make fun of weather persons telling us that there is a 40% chance of rain next Tuesday, but that prediction is based on a much better knowledge base than, say, any claim made by a hedge fund manager or any claim by a general that we'll have accomplished our missions in Afghanistan and Iraq by...(pick any date you want; I guarantee it will be wrong).

In earlier chapters we discussed several ways that causal reasoning causes many people all kinds of problems. For example, no matter how many excellent scientific studies show that acupuncture has no intrinsic clinical value and is a form of placebo medicine, there will always be someone who thinks those studies must be wrong because "acupuncture is the only thing that helps my migraines" or some such thing. (See Appendix C.) The *illusion of causality* gets full coverage in *The Invisible Gorilla*. There's the background involving our evolutionary history that has produced a species with stupendous pattern-recognition abilities, so stupendous, in fact, that we often see patterns where there are none. We've evolved to find meaning in patterns and infer causal relationships from coincidences. "Our understanding of our world is systematically biased to perceive meaning rather than randomness and to infer cause rather than coincidence. And we are usually completely unaware of these biases." Consider how automatic some of these brain processes are:

In fact, visual areas of your brain can be activated by images that only vaguely resemble what they're tuned for. In just one-fifth of a second, your brain can distinguish a face from other objects like chairs or cars. In just an instant more, your brain can distinguish objects that look a bit like faces, such as a parking meter or a three-prong outlet, from other objects like chairs. Seeing objects that resemble faces induces activity in a brain area called the fusiform gyrus that is highly sensitive to real faces. In other words, almost immediately after you see an object that looks anything like a face, your brain treats it like a face and processes it differently than other objects. (Chabris and Simons, *The Invisible Gorilla*)

Add a little religious or political zealotry to the brain's natural disposition to recognize faces in just about anything with a shape and a few shadows and you've got the recipe for a dozen tortillas with the "face of Jesus" imprinted on them or a single toasted cheese sandwich that reminds people of President Obama or Madonna.

The reader is probably tired of my mantra that one of the hardest lessons to learn is that personal experience is not always the best guide to what's true or to what's even *relevant* to what's true. Yet, I can't resist one more example. Just because you know that you get aches and pains according to what kind of weather you're experiencing, doesn't mean you're right. Telling people who believe they are human barometers that scientific studies haven't found any consistent connection between changes in the weather and changes in people's aches and pains isn't likely to convince them that they're deluding themselves. They've experienced it, and that's that. No amount of discourse on the *post hoc fallacy*, the regressive fallacy, confirmation bias, conditioning, or the power of suggestion will change their minds. It's almost as if there's a conspiracy in nature to lead people into error. Actually, there is.

We've already discussed how the brain constructs perceptions out of bits and pieces of sense data. The brain also constructs causal narratives out of bits and pieces of data. Unfortunately, many people look for a single cause or reason for complex events, which misleads them even further. Add a celebrity or two, a doctor in a white coat, and a push from the media, and any number of erroneous causal beliefs can become widespread. The anti-vaccination movement and the popularity of "alternative" cancer cures exemplify how this works. Even those of us who know better have found again and again how hard it is to make ourselves try to falsify beliefs. It is not natural to look for counter-stories where disease didn't follow a breast implant or where autism didn't follow a vaccination. At the very least, one would hope that by becoming aware of the many ways our brain can trick us, we would arrive at the conclusion Bertrand Russell thought was a necessary consequence of the limits of knowledge: we should be less cocksure of our beliefs, hold them tentatively, and always be on guard against thinking our feeling of absolute certainty implies we're right.

We live in a world where intuition is valued more than science by many people, where absurdities like homeopathy and horoscopes do not decrease in popularity despite the widely available and easily accessible evidence that they are nonsense. We live in a world where entertainment has taken a decidedly paranormal turn toward ghosts and spiritualism and psychic baloney. The really good news for those of us who live in the free world is that we are free to *criticize or advocate* astrology, anti-scientific medicine, psychic mediumship, and other forms of superstition. The defenders of critical thinking, skepticism, and science should not whine about the fact that our arguments don't persuade large numbers of people. As long as the astrologers and spiritualists are not in power and not writing laws that force us to think like them, we shouldn't complain about the effectiveness of our movement to spread the good news about science and critical thinking. We have to accept that our liberty depends on their freedom to believe what they want, no matter how unscientific or false. Toleration of the wrong and the misguided is a price we must pay for our liberty. (I'm not so naïve as to think that we in the free world are completely free. I've watched with the rest of you as a president of the United States, with bipartisan support, claimed to be defending freedom and liberty while he was secretly approving subpoenas, arrests, detentions, and trials for those secretly identified as "suspected terrorists." I also remember the "Imperial Presidency" of Richard Nixon with his administration's enemies list whose purpose was so eloquently described by John Dean as to "use the available federal machinery to screw our political enemies.") Anyway, your neighbor who has been trying to get you to take *Oscillococcinum* during flu season will not be taken by ambulance to a homeopathic clinic if she calls 9-1-1. In the U.S., prospective employers aren't going to ask you what your Sun sign is or have your handwriting analyzed by a graphologist before they will consider hiring you. So far, no medium has been allowed in court to summon the dead to testify

I realize that there are many politicians—most of them Republicans, I must point out—who do not accept evolution or other basic scientific facts. Many politicians—no matter what their party loyalty—claim that they pray for guidance from a god to help them make important decisions that affect all of us. This is troubling. Some even claim that their decision to run for president of the United States was inspired by a message from a god. This is very troubling. There is no compelling evidence that gods of any sort pay any attention to human behavior. Yet, such belief seems natural to many humans. So does the equally troubling promotion of the idea that following one's intuitions, instincts, or gut feelings needs to be cultivated and nourished. We've had a president (George W. Bush) who openly admitted that he follows his gut when making important decisions, including—presumably—the decision to send men and women off to battle. Oprah Winfrey—arguably the most popular television star of all time—repeatedly promoted the idea that intuition should be trusted above science. The fact is that it is our natural inclination to go with our gut feelings and intuitions. We don't need to cultivate this tendency. What we need to cultivate is some skepticism about the reliability of our instincts and immediate interpretations of our perceptions. Our instincts obviously serve us well; otherwise, our species would have gone extinct a long time ago. Questioning our instincts, gut feelings, first impressions, and what the majority of people around us believe does not come naturally. We must consciously develop the habit of looking to the science in a fair and unbiased way to check our instincts. We must work hard at overcoming the natural tendency to see causal connections and other patterns where there are none by examining *all* the evidence, not just the data that support our gut feeling. Critical thinking, skepticism, and science did not evolve on the savannah millions of years ago. They are unnatural and go against the grain of those instincts that helped our species survive for hundreds of thousands of years.

Cultivating Your Unnatural Self

If you are willing to be open-minded, accept that reasonable probabilities rather than absolute certainties are the best you can get in many of the things that matter, and hold your most precious beliefs tentatively (always being willing to reconsider your beliefs and the reasons that led you to them), then there is hope that you will overcome some of the hindrances to critical thinking at least some of the time. If you can avoid the errors and apply the skills discussed in previous chapters, there is hope that at least occasionally you will hit the mark. Finally, if you will become a goal-oriented thinker—setting goals for yourself to improve your critical thinking ability and setting specific goals when facing a problem or decision—you may find that critical thinking is not only possible but is as rewarding as any unnatural act you've ever engaged in. Well, maybe that's a bit of an exaggeration. In any case, the most important goal you can set as a thinker is the goal to get it right, to be accurate. But achieving that goal is hindered by many natural and cultivated tendencies. It will do you little good, for example, to seek accuracy if the methods you select for getting and evaluating information are biased. A physician may be motivated to make a correct diagnosis of a patient, but if she bases her diagnosis on the similarity of symptoms to several recently examined patients, she may be forsaking accuracy for availability. Dr. Jerome Groopman (2007) gives the example of a doctor who had treated scores of patients over a period of several weeks for viral pneumonia. A patient then came in with similar symptoms, except that her chest X-ray didn't have the characteristic white streaks of viral pneumonia. The doctor diagnosed her as being in the early stages of the illness. He was wrong. Another doctor diagnosed her correctly as suffering from aspirin toxicity. The diagnosis of viral pneumonia was available because of the recent experience of many cases of the illness. After he realized his mistake, he said "it was an absolutely classic case—the rapid breathing, the shift in her blood electrolytes—and I missed it. I got cavalier."

The *availability error* is just one of many tendencies that can hinder our achievement of accuracy in thinking. Desiring to be accurate will not be enough to achieve our goal. We must also use the best methods of thinking available to us. Many of these go against the grain of our natural tendencies. To help you in your unnatural endeavors, I will conclude with a chapter on 59+ Ways to Develop Your Unnatural Talents in Critical Thinking, Skepticism, and Science.

SOURCES FOR CHAPTER NINE 276

X
59+ Ways to Develop Your Unnatural Talents in Critical Thinking, Skepticism, and Science

"The rule that human beings seem to follow is to engage the brain only when all else fails — and usually not even then."-- David Hull

"Most people would die sooner than think — in fact they do so."—Bertrand Russell

How the Brain Works: Some Sources

Learn as much as you can about how the brain works. It is especially important to learn how memory and sense perception work, and how the brain tricks and deceives us—usually for our own good. There are many excellent books available to help you with this task. See, for example *Phantoms in the Brain: Probing the Mysteries of the Human Mind* and *The Tell-Tale Brain: A Neuroscientist's Quest for What Makes Us Human* by V. S. Ramachandran; any book by Oliver Sacks, Daniel L. Schacter, or Elizabeth Loftus; *A Mind of Its Own: How Your Brain Distorts and Deceives* by Cordelia Fine; *SuperSense: Why We Believe in the Unbelievable* by Bruce Hood; or any of the books listed in sources for Chapter Nine. I don't usually recommend books I haven't read, but Daniel Kahneman's *Thinking, Fast and Slow* was published as I was writing this last chapter. Kahneman is a Nobel Prize winner in economics and is known for identifying the many ways we are irrational in our judgments and decision making. I have relied on earlier work by Kahneman and his co-author Amos Tversky for several of the 59 items listed below. Whenever I think of Kahneman I think of the following test: "A bat and a ball cost $1.10 in total. The bat costs $1 more than the ball. How much does the ball cost?" If you understand the significance of this simple test, you understand the core message of *Unnatural Acts*. Another book I have on my reading list came out shortly after Kahneman's: *The Folly of Fools: The Logic of Deceit and Self-Deception in*

Human Life by Robert Trivers. Finally, a book I think will enjoy is Al Seckel's *Incredible Visual Illusions:)* *Believe Your Eyes.*

The 59 Ways

Set aside some time each day to investigate further the following 59 affective, cognitive, or perceptual biases, fallacies, or illusions. Some of these topics—those listed in **boldface**—are discussed in previous chapters or in one of the appendices. A short description of topics not discussed in previous chapters is given to get you started in your research. Two online sources I recommend are Wikipedia <en.wikipedia.org/wiki/Main_Page> and The Skeptic's Dictionary <**www.skepdic.com**>. A good place to start in Wikpedia is with the entry that lists cognitive biases <en.wikipedia.org/wiki/Cognitive_biases >. A good place to start in The Skeptic's Dictionary is with the entry on hidden persuaders <skepdic.com/hiddenpersuaders.html>. Follow my blog at **59ways.blogspot.com** to help you in your research. You will soon discover that the 59 items listed here represent the tip of the iceberg of biases and illusions.

1. *ad hoc* **hypothesis**
2. *ad hominem*
3. *ad populum*
4. affect bias
   Our judgment regarding the costs and benefits of items is often significantly influenced by a feeling evoked by pictures or words not directly relevant to the actual cost or benefit. For some, the good or bad feeling they have just prior to making a decision is a bias that influences that decision and renders it irrational.
5. anchoring effect
   Our judgment regarding the frequency, probability, or value of items is often determined by comparing the item to an anchor point. For some the anchor is a bias that determines many of their decisions. For example, some surgeons always recommend spinal fusion for

lower back pain; some therapists diagnose most of their patients with multiple personality disorder, while other therapists never see a case of MPD in their entire career. If the label on a coat in a clothing store has a price tag with three different prices, the two highest of which are crossed out, you may think you are getting a bargain if you accept the highest price on the tag as an anchor.

6. apophenia
   Apophenia is the spontaneous perception of connections and meaningfulness of unrelated phenomena. For example, a man's son killed himself and the father spontaneously saw a stopped clock as a sign from his son, informing the father of the time of death and validating his belief in the afterlife.

7. **appeal to authority**
8. **appeal to tradition**
9. **argument to ignorance (*argumentum ad ignorantiam*)**
10. autokinetic effect
    The autokinetic effect refers to perceiving a stationary point of light in the dark as moving.

11. **availability error**
12. Barnum effect
    The Barnum effect is the name given to a type of **subjective validation** in which a person finds personal meaning in statements that could apply to many people.

13. **backfire effect**
14. begging the question
    Begging the question is what one does in an argument when one assumes what one claims to be proving. Here's an example: "Past-life memories of children prove that past lives exist because the children could have no other source for their memories besides having lived in the past."

15. **change blindness**
16. **classical conditioning**

17. Clever Hans effect
18. clustering illusion
19. cognitive dissonance

Cognitive dissonance is a theory of human motivation that asserts that it is psychologically uncomfortable to hold contradictory cognitions. The theory is that dissonance, being unpleasant, motivates a person to change his cognition, attitude, or behavior.

20. **coincidence (law of truly large numbers)**
21. cold reading

Cold reading refers to a set of techniques used by professional manipulators to get a subject to behave in a certain way or to think that the cold reader has some sort of special ability that allows him to mysteriously know things about the subject. Cold reading goes beyond the usual tools of manipulation: suggestion and flattery. The cold reader banks on the subject's inclination to find more meaning in a situation than there actually is. The manipulator knows that his mark will be inclined to try to make sense out of whatever he is told, no matter how farfetched or improbable. He knows, too, that people are generally self-centered, that we tend to have unrealistic views of ourselves, and that we will generally accept claims about ourselves that reflect not how we are or even how we think we are but how we wish we were or think we should be. He also knows that for every several claims he makes about you that you reject as being inaccurate, he will make one that meets with your approval; and he knows that you are likely to remember the hits he makes and forget the misses.

22. communal reinforcement
23. confabulation
24. confirmation bias
25. continued influence effect
26. false dichotomy

210

27. false implication
    The fallacy of false implication occurs when a statement, which may be clear and even true, implies that something else is true or false when it isn't. For example, if I write in my 30-day evaluation log of an employee that on May 15th she was on time for work, someone reading the log might infer that this was unusual and that usually the employee did not arrive on time. Perhaps she is always on time but by indicating her promptness just once I can give the false impression that she is usually late for work.

28. false memories

29. file-drawer effect

30. Forer effect
    The Forer effect refers to the tendency of people to rate sets of statements as highly accurate for them personally even though the statements could apply to many people.

31. gambler's fallacy
    The gambler's fallacy is the mistaken notion that the odds for something with a fixed probability increase or decrease depending on recent occurrences.

32. halo effect
    The halo effect refers to a bias whereby the perception of a positive trait in a person or product positively influences further judgments about traits of that person or products by the same manufacturer. One of the more common halo effects is the judgment that a good looking person is intelligent and amiable. There is also a reverse halo effect whereby perception of a negative or undesirable trait in individuals, brands, or other things influences further negative judgments about the traits of that individual, brand, etc.

33. hindsight bias

34. hypersensory perception
    Hypersensory perception is a term coined by Theodore Schick, Jr. and Lewis Vaughn to describe what some people mistakenly call intuition. A person with HSP is

very observant and perceptive, and may appear to be psychic.

35. ideomotor effect
36. inattentional blindness
37. magical thinking
38. mere exposure effect

The mere exposure effect is a tendency to favor or feel more comfortable with a person or situation only because you've been exposed to that person or situation before.

39. motivated reasoning
40. non sequitur
41. pareidolia

Pareidolia is a type of illusion or misperception involving a vague or obscure stimulus being perceived as something clear and distinct. For example, in the discolorations of a burnt tortilla one sees the face of Jesus. Or one sees the image of Mother Teresa or Ronald Reagan in a cinnamon bun, or Satan in rain drip patterns on a patio umbrella or soap stains on a shower curtain.

42. placebo effect
43. positive-outcome bias

Positive-outcome bias (or "publication bias") is the tendency to publish research with a positive outcome more frequently than research with a negative outcome. Positive-outcome bias also refers to the tendency of the media to publish medical study stories with positive outcomes much more frequently than such stories with negative outcomes. Negative outcome refers to finding nothing of statistical significance or causal consequence, not to finding that something affects us negatively.

44. post hoc fallacy
45. priming
46. regressive fallacy

The regressive fallacy is the failure to take into account natural and inevitable fluctuations of things when ascribing causes to them. Things like stock market prices, golf scores, and chronic back pain inevitably fluctuate. Periods of low prices, low scores, and little or no pain are eventually followed by periods of higher prices, scores, pain, etc.

47. representativeness error

In judging items, we compare them to a representative idea and tend to see them as typical or atypical according to how they match up with our model. Jerome Groopman, M.D., gives the example of a doctor who failed to diagnose a cardiac problem with a patient because the patient did not fit the model of a person likely to have a heart attack. The patient complained of all the things a person with angina would complain of, but he was the picture of health. He was in his forties, fit, trim, athletic, worked outdoors, didn't smoke, and had no family history of heart attack, stroke, or diabetes. The doctor wrote off the chest pains the patient complained of as due to overexertion. The next day the patient had a heart attack.

48. retrospective falsification

D. H. Rawcliffe coined this term to refer to the process of telling a story that is factual to some extent, but which gets distorted and falsified over time by retelling it with embellishments. The embellishments may include speculations, conflating events that occurred at different times or in different places, and the incorporation of material without regard for accuracy or plausibility. The overriding force that drives the story is to find or invent details that fit with a desired outcome. The process can be conscious or unconscious. The original story gets remodeled with favorable points being emphasized and unfavorable ones being dropped. The distorted and false version becomes a memory and record of a remarkable tale.

49. selection bias

Selection bias comes in two flavors: (1) self-selection of individuals to participate in an activity or survey, or as a subject in an experimental study; (2) selection of samples or studies by researchers to support a particular hypothesis.

50. selective thinking

51. self-deception

52. shoehorning

Shoehorning is the process of force-fitting some current affair into one's personal, political, or religious agenda. People claiming to be psychic frequently shoehorn events to fit vague statements they made in the past. This is an extremely safe procedure, since they can't be proven wrong and many people aren't aware of how easy it is to make something look like confirmation of a claim after the fact, especially if you give them wide latitude in making the shoe fit.

53. straw man

54. subjective validation

55. sunk-cost fallacy

When one makes a hopeless investment, one sometimes reasons: I can't stop now, otherwise what I've invested so far will be lost. This is true, of course, but irrelevant to whether one should continue to invest in the project. Everything one has invested is lost regardless. If there is no hope for success in the future from the investment, then the fact that one has already lost a bundle should lead one to the conclusion that the rational thing to do is to withdraw from the project.

56. suppressed evidence

One of the basic principles of cogent argumentation is that a cogent argument presents all the relevant evidence. An argument that omits relevant evidence appears stronger and more cogent than it is. The fallacy of suppressed evidence occurs when an arguer intentionally omits relevant data. This is a difficult fallacy to detect because we often have no way of knowing that we haven't been told the whole truth.

57. testimonials (anecdotal evidence)

58. Texas-sharpshooter fallacy

The Texas-sharpshooter fallacy is the name epidemiologists give to the **clustering illusion**. Politicians, lawyers, and some scientists tend to isolate clusters of diseases from their context, thereby giving the illusion of a causal connection between some environmental factor and the disease. What appears to be statistically significant (i.e., not due to chance) is actually expected by the laws of chance.

59. wishful thinking

Wishful thinking is interpreting facts, reports, events, perceptions, etc., according to what one would like to be the case rather than according to the actual evidence.

Some Science Sources

There are many excellent books that introduce the reader to the history, methods, and accomplishments of science. A few that I have found interesting and useful are, in no particular order: Bill Bryson's *A Short History of Nearly Everything*; Jacob Bronowski's *The Ascent of Man*; any book by Richard Dawkins, Stephen Jay Gould, or Carl Sagan; Donald Prothero's *Evolution: What the Fossils Say and Why It Matters*; Phil Plait's *Bad Astronomy: Misconceptions and Misuses Revealed, from Astrology to the Moon Landing "Hoax"*; and Timothy Ferris's *The Whole Shebang: A State-Of-The-Universe's Report*.

Afterword

"The church teaches us that we can make God happy by being miserable ourselves."—Robert Ingersoll

---

I noted above that when I refer to an idea as unworthy of consideration (or words to that effect that might sound a bit harsher in context) I do not intend to cast aspersions on the intelligence or character of the one holding the unworthy belief. I must qualify that disclaimer. When I see or hear of anyone murdering strangers because they believe such acts please their god, I consider not only the idea but the one holding such an idea to be unworthy of any sympathetic consideration. How anyone could come up with the idea that an almighty being would be pleased by causing as much misery as possible escapes my imagination. Such a person also escapes my capacity for mercy.

Appendix A
Cell phones, Radiation, and Cancer

The Science Says No

Many people fear that electromagnetic fields (EMFs) cause cancer; however, a causal connection between EMFs and cancer has not been established. The National Research Council (NRC) spent more than three years reviewing more than 500 scientific studies that had been conducted over a 20-year period and found "no conclusive and consistent evidence" that electromagnetic fields harm humans. The chairman of the NRC panel, neurobiologist Dr. Charles F. Stevens, said that "Research has not shown in any convincing way that electromagnetic fields common in homes can cause health problems, and extensive laboratory tests have not shown that EMFs can damage the cell in a way that is harmful to human health."

In 1997, *The New England Journal of Medicine* published the results of the largest, most detailed study of the relationship between EMFs and cancer ever done. Dr. Martha S. Linet, director of the study, said: "We found no evidence that magnetic field levels in the home increased the risk for childhood leukemia." The study took eight years and involved measuring the exposure to magnetic fields generated by nearby power lines. A group of 638 children under age fifteen with acute lymphoblastic leukemia were compared to a group of 620 healthy children. "The researchers measured magnetic fields in all the houses where the children had lived for five years before the discovery of their cancer, as well as in the homes where their mothers lived while pregnant." The study was criticized because it is impossible to know exactly what the EMFs were at the times the mothers or their children were exposed. All measurements must be done after the exposure has taken place and assumptions must be made that the level of EMFs was not substantially different during exposure

A report published in the *Journal of the American Medical Association* on a study of 891 adults who used their cell phones between 1994 and 1998 found that there was no increased risk of brain cancer associated with cell phone use (Muscat 2000).

Scientific studies are not all of equal value or significance. Studies vary in quality, in the number of data points, and in duration. Some study a few people for a short time. Even the best designed and executed of such studies can't be any more than suggestive. Epidemiological studies, by their very nature, cannot establish a positive causal link, but they can indicate a high probability of there *not* being a causal link if no correlation is found between two studied phenomena. A large Danish study (420,000 mobile phone users), for example, found neither long nor short-term mobile phone use associated with an increased risk of cancer (2006 and 2011). The Danish study did not use the memories of the subjects to assess exposure; they analyzed data from mobile phone company records. Another small Swedish study found no increased risk of acoustic neuroma related to short-term mobile phone use (2004). The researchers thought that their data suggest an increased risk of acoustic neuroma associated with mobile phone use of at least 10 years duration. They don't say how they measured exposure, but they note that "detailed information about mobile phone use and other environmental exposures was collected." Other studies on laboratory animals have found effects from microwave exposure. Some studies have collected data suggestive of possible harmful effects from cell phone microwave exposure, but they are too small to have ruled out chance or other causal agents or they have not been tested on in vivo cells.

Photons and Non-ionizing Radiation

All electromagnetic radiation comes from photons. The energy of a photon depends on its frequency. Cellular phones operate at the radio frequency (RF) part of the electromagnetic spectrum. This is non-ionizing radiation. Other examples of the non-ionizing part of the electromagnetic spectrum include AM and FM radio waves, microwaves, and infrared waves from heat lamps. Unlike x-rays and gamma rays (which are examples of ionizing radiation), radio waves have too little energy to break the bonds that hold molecules (such as DNA) in cells together. Similarly, since RF of this frequency contains relatively low energy, it does not enter tissues. At very high levels of exposure, RF can cause warming of tissues, much as a heat lamp does. The wavelength of cell phone waves is about one foot and the frequency is approximately 800 to 900 MHz, although newer models may use higher frequencies up to 2,200 MHz.

"Roughly one million photons in a power line together have the same energy as a single photon in a microwave oven, and a thousand microwave photons have the energy equal to one photon of visible light" (Lakshmikumar 2009). Ionizing radiation is known to cause health effects; "it can break the electron bonds that hold molecules like DNA together" (Trottier 2009). "The photon energy of a cell phone EMF is more than 10 million times weaker than the lowest energy ionizing radiation" (Trottier 2009). Thus, the likelihood that our cell phones, microwave ovens, computers, and other electronic devices are carcinogenic is miniscule.

The Demand for the Impossible Study

Nevertheless, it is impossible to prove that no study will ever find a significant correlation between EMFs and cancer or any other disease or disorder. No product can be shown to be absolutely safe for everybody. For example, a 2004 *British Medical Journal* article claimed to have found an inexplicable increase in leukemia in children living near power lines in England and Wales. The researchers wrote: "There is no accepted biological mechanism to explain the epidemiological results; indeed, the relation may be due to chance or confounding." On the other hand, a 2003 study of women on Long Island found no causal connection between living near power lines and developing breast cancer. A single study does not prove there is or there isn't a causal link between EMFs and cancer. We have to look at what is indicated by the preponderance of the evidence from all the studies.

There is also a strong contingent of folks hell-bent on proving this link, so it is likely that studies will continue to be done that support a contrary viewpoint. For example, a research team in Sweden found an increased risk for brain tumors in people who used cellular or cordless phones (2006). The study was a small one and assessed exposure by self-administered questionnaires. The media and fear mongers made a fuss over a report issued by the World Health Organization (WHO) in June 2011 from a group of thirty-one scientists from fourteen countries. The group is known as the International Agency for Research on Cancer (IARC). Jonathan Samet, who headed the group, said that "After reviewing essentially all the evidence that is relevant ... the working group classified radiofrequency electromagnetic fields as *possibly carcinogenic to humans*" (emphasis added). Samet said that *some evidence suggested* a link between an increased risk for glioma, a type of brain cancer, and mobile phone use. The WHO panel justified classifying cell phones as a *possible* carcinogen by selecting one bit of one study from among dozens of studies that were part of a group of studies known as the Interphone study. That one study was published in September 2010 in the *International Journal of Epidemiology*. The conclusion of the scientists who published the study was:

Overall, no increase in risk of glioma [malignant brain tumor] or meningioma [benign brain tumor] was observed with use of mobile phones. *There were suggestions of an increased risk of glioma at the highest exposure levels, but biases and error prevent a causal interpretation.* The possible effects of long-term heavy use of mobile phones require further investigation. [emphasis added]

Disconnected Media

A year before the 2011 IARC report, fears about cell phones and cancer were aroused to fever pitch by Devra Davis in her book *Disconnect: The Truth About Cell Phone Radiation, What the Industry Has Done to Hide It, and How to Protect Your Family*. The "disconnect" that Davis, an epidemiologist, highlights in her opening salvo against the cell phone industry is revealed in her first anecdote. She tells the story of a woman who gets a headache in a meeting and suspects that somebody in the room has left his cell phone on. Who hasn't been in a group where somebody's left his cell phone on? Anyway, the woman is right. Somebody has his cell phone on and he turns it off. We're not told whether the woman's headache turned off with cell phone disconnect. We are told that the woman is Gro Harlem Brundtland, a physician who used to be the prime minister of Norway. She also served as director general of the World Health Organization and while there banned the use of cell phones in her office. The "disconnect" in the title of the book is revealed when the author notes that after many years of study and work the WHO's report on cell phones was about to be released and concluded that further study was needed. Why? The "disconnect" is that the data didn't support the beliefs of Gro Harlem Brundtland and Devra Davis. They are certain that cell phones cause headaches and cancers, despite what the data say. Perhaps their spirits were lifted when the IARC report included the claim that there's a *possible* connection between EMFs and cancer.

Davis's book is just one of several appealing to fear and anecdotes combined with the can't-miss appeal to an industry cover-up and conspiracy to hide the real data. In 2009 Carleigh Cooper came out with *Cell Phones and The Dark Deception: Find Out What You're Not Being Told...And Why*. Christopher Ketcham also unfolded a fable in GQ (February 2010) about a government/industry conspiracy and cover-up. Ketcham's scare piece, entitled "Warning: Your Cell Phone May Be Hazardous to Your Health," featured a graphic of a cell phone placed next to a package of Marlboro cigarettes. Graphics like this always help the slower reader who might miss the point that Ketcham would make. I will admit that he puts forth a convincing case that cell phones and Wi-Fi are harmful, that the government knows this, the industry knows this, and both have actively silenced those who say otherwise. The only ones who have been telling the truth about the dangers of cell phones are mass media newspapers and magazines, a guy named Allan Frey who hasn't done any research for twenty-five years, and a guy named Louis Slesin who's made a living out of scaring people about EMFs since 1980. All the science that has found no causal connection between cell phones, cell phone towers, Wi-Fi, and Wi-Fi towers is tainted and biased because of its connection to the industries that are making billions on this stuff. Ketcham reminds the reader that the same thing happened with tobacco and asbestos. The only thing missing from Ketcham's story, however, is the *evidence* similar to what we have for the tobacco and asbestos stories. Like Davis, he makes his case seem stronger than it is by being selective in his presentation of data and by giving more weight to small and insignificant studies than they deserve. He relies heavily on the Interphone studies, but he doesn't tell the reader that none of the findings in the Interphone studies show a robust correlation between cellphone usage and brain tumors or other diseases. No causal connections have been established at all. Where positive correlations were found, the authors used cautious language, e.g., "possibly reflecting participation bias or other methodological limitations," "finding could either be causal or artifactual, related to differential recall between cases and controls," "based on few subjects (7 cases and 4 controls) needs to be investigated further," and "additional investigations of this association...are needed to confirm these findings." It is important when reporting on a scientific study to provide these kinds of details. Otherwise, the writer distorts the science, arousing false hope or fear based on inconclusive data. This kind of incomplete, misleading reporting is typical of the mass media's treatment of scientific studies. It is also revealing that most of Ketcham's sources are mass media reports, not scientific studies. Also, he doesn't mention that the WHO sponsored the Interphone studies, perhaps because he later attacks the WHO and claims that the cell-phone industry has influenced the WHO by making payments "to WHO personnel working on wireless health effects."

The lead-in to Ketcham's article sets the tone for the article itself: it aims to frighten the reader by suggesting two untruths: that the evidence is piling up that cell phones are damaging brains and that there is a cover-up.

> Ever worry that that gadget you spend hours holding next to your head might be damaging your brain? Well, the evidence is starting to pour in, and it's not pretty. So why isn't anyone in America doing anything about it?

The evidence is piling up, but it is piling up in the other direction: there is growing evidence that cell phones are *not* damaging our brains. Furthermore, even though there is no need for it, there are government officials who are trying to spread the word about the dangers of cell phones. State Rep. Andrea Boland of Maine believes that "numerous studies point to the cancer risk." She proposed that a law be passed in her state that would require cell phone manufacturers to put warnings on packaging like those on cigarettes. Apparently Boland reads the same newspapers that Ketcham reads. She, too, cites the Interphone Studies without mentioning the cautions that the authors of those studies have stated about the results of their research. Boland relies on a retired electronic engineer, Lloyd Morgan, for much of her support. Boland's bill failed to pass by a margin of 83-62.

Ketcham begins his scare piece with a story about an anonymous guy who, along with his anonymous doctor, thinks that his tumor was caused by his cell phone and that there is a conspiracy by the cell phone industry to discredit studies that show cell phones are dangerous. Ketcham's hook to the dangerous conspiracy theme is to cite scary headlines in mass media publications from around the globe. He then asserts that "the scientific debate is heated and far from resolved. There are multiple reports, mostly out of Europe's premier research institutions, of cell-phone and PDA use being linked to "brain aging, brain damage, early-onset Alzheimer's , senility, DNA damage, and even sperm die-offs." Yes, studies have *linked* cell phones to all kinds of disorders, but links are *correlations* and are not necessarily indicative of a causal connection. Links might be found that are due to chance or to some artifact of the method used, such as having a small or biased sample. It is true that some studies have collected data suggestive of possible harmful effects from cell phone microwave exposure, but they are too small to have ruled out chance or other causal agents or they have not been tested on in vivo cells. The data do not support Ketcham's fears. If Ketcham has looked at these reports, he doesn't reveal it. Instead, he simply notes that in September 2007 the European Environment Agency warned that cell-phone technology "*could* lead to a health crisis similar to those caused by asbestos, smoking, and lead in petrol" (emphasis added). True enough. The EEA did suggest that there *could* be dangers from cell phones. It also noted that just because the data so far do not indicate a serious health threat does not mean that at some time in the future we won't find out otherwise. True enough, but we have a vast amount of data on cell phone usage that strongly indicates the fear of brain cancer from using cell phones is unwarranted.

Ketcham's two main sources for his article are Allan Frey and Louis Slesin. Frey hasn't done any research for twenty-five years. Frey was a neuroscientist who worked on biological effects of microwaves. Ketcham writes:

> In a study published in 1975 in the *Annals of the New York Academy of Sciences*, Frey reported that microwaves pulsed at certain modulations could induce "leakage" in the barrier between the circulatory system and the brain. Breaching the blood-brain barrier is a serious matter: It means the brain's environment, which needs to be extremely stable for nerve cells to function properly, can be perturbed in all kinds of dangerous ways. Frey's method was rather simple: He injected a fluorescent dye into the circulatory system of white rats, and then swept the microwave frequencies across their bodies. In a matter of minutes, the dye had leached into the confines of the rats' brains.
>
> Frey says his work on radar microwaves and the blood-brain barrier soon came under assault from the government. Scientists hired and funded by the Pentagon claimed they'd failed to replicate his findings, yet they also refused to share the data or methodology behind their research ("a most unusual action in science," Frey wrote at the time). For more than fifteen years, Frey had received almost unrestricted funding from the Office of Naval Research. Now he was told to conceal his blood-brain-barrier work or his contract would be canceled.

Whatever the truth about Frey and the government, it is a fact that work in this area has continued. Microwave ray guns are being used to control crowds. We know that microwaves can be harmful to our brains. Ketcham makes no effort to sort out the differences between what's going on in a microwave oven or a microwave weapon from what's going on with a cell phone or with radiation from sunlight for that matter.

Unlike Frey, Slesin is not a trained scientist. He has a doctorate in environmental policy from MIT and has been publishing *Microwave News* since 1980. He has rejected all the science done in the past 30 years that has found no support for the hypothesis that cell phones and Wi-Fi are hazardous to our health. He makes a living out of warning people of the dangers of microwaves. As scientific studies got larger and better designed, the EMF-cancer connection grew weaker. As the evidence has continued to pile up against the EMF-cancer connection, Slesin has dug in deeper, rather than admit that he was wrong. Physicist Bob Park wrote about Slesin in *Voodoo Science*. Park notes that in 1996 "the National Academy of Sciences released the results of an exhaustive three-year review of possible health effects from exposure to residential electromagnetic fields."

> The large conference room in the Academy building on Constitution Avenue in Washington was crowded with reporters, TV cameras, and a few scientists. The head of the review panel, Charles Stevens, a distinguished neurobiologist with the Salk Institute, summed up the results: "Our committee evaluated over 500 studies, and in the end all we can say is that the evidence doesn't point to these fields as being a health risk."
>
> There were reporters in the room who had been writing stories about the dangers of power-line fields for years. For Louis Slesin, the editor of *Microwave News*, an influential newsletter devoted entirely to the EMF health issue, the controversy was his livelihood. For these reporters to now write that it had all been a false alarm would have been miraculous. They would scour the report looking for soft spots. But the evidence against a connection between electricity and cancer was getting harder to ignore.

Rather than admit he was wrong, Slesin told Ketcham: "We love our cell phones. The paradigm that there's no danger here is part of a worldview that had to be put into place. Americans are not asking the questions, maybe because they don't want the answers. So what will it take?" According to Ketcham, Frey has the answer to Ketcham's question: "Until there are bodies in the streets," he said, "I don't think anything is going to change." That's how Ketcham ends his article. To call this shoddy journalism is an insult to shoddy journalism.

Ketcham also lionizes the work of Olle Johansson, Ph.D., a Swedish neuroscientist and ardent anti-EMF crusader who was awarded the not-so-prestigious Misleader of the Year award in 2004 by the Swedish Sceptics (Vetenskap och Folkbildning, VoF). He was given his title "Following many years of public assertions and cocksure, blatant warnings of numerous negative health effects allegedly caused by electromagnetic fields." Despite growing evidence against his position, Johansson continues to publish reports alleging a variety of unpleasant health conditions, diseases, and disorders are due to electromagnetic radiation. He has staked his career on this position, despite its conflicting with the consensus of the scientific evidence and community. A number of rigorous studies have been done on the health dangers of EMFs and Johansson wasn't involved in any of them. Johansson is not a physicist, but a neuroscientist, yet he makes claims as if he were an expert in physics. He compares microwaves with X-rays and gamma radiation, although these different sorts of electromagnetic waves relate to entirely different physical phenomena.

For some reason, however, none of those hyping the dangers of cell phones mentioned that one study found that cell phone radiation boosts memory and reverses Alzheimer's....in rats.

Wi-Fi Alarms

Ketcham also jumps on the bandwagon of those warning us about the dangers of damage to our brains from exposure to Wi-Fi. Ketcham notes that Wi-Fi operates at the same frequency as microwave ovens (about 2.4 gigahertz), but he doesn't mention that Wi-Fi uses very low intensity radio waves. Nor does Ketcham tell the reader that Wi-Fi radiation is 100,000 times less than that of a domestic microwave oven and the power levels for Wi-Fi are lower than that for cell phones. Nor does Ketcham note that the modulated frequencies bringing radio and television transmissions into our homes are stronger and more pervasive than the radio waves used by wireless networks.

Ketcham mentions a rural town in Sweden where residents claimed they were getting sick because of Wi-Fi towers, but he doesn't mention a similar case in South Africa. Protestors handed out flyers warning residents of Craigavon, outside of Johannesburg, South Africa, that microwaves from a newly erected cellphone tower would cause health problems. Soon, residents complained of rashes, headaches, nausea, tinnitus, dry burning itchy skins, gastric imbalances, and totally disrupted sleep patterns. The only problem was that even when the tower was turned off for six weeks (unbeknownst to the residents), the residents still complained of their many ailments.

One university president has banned Wi-Fi on his campus, claiming that microwave radiation in the frequency range of Wi-Fi has been shown to increase permeability of the blood-brain barrier, cause behavioral changes, alter cognitive functions, activate a stress response, interfere with brain waves, cell growth, cell communication, calcium ion balance, etc., and cause single and double strand DNA breaks. What compelling scientific evidence he has for making such claims is not known.

Another school banned Wi-Fi after a classics teacher complained that it was making him physically ill. The teacher said that after Wi-Fi was in place:

> I felt a steadily widening range of unpleasant effects whenever I was in the classroom. First came a thick headache, then pains throughout the body, sudden flushes, pressure behind the eyes, sudden skin pains and burning sensations, along with bouts of nausea. Over the weekend, away from the classroom, I felt completely normal.

That the teacher's symptoms were caused by Wi-Fi is, however, pure speculation and unlikely. There is no compelling scientific evidence that these kinds of symptoms are caused by exposure to Wi-Fi.

Electrosensitivity and Mass Media Scare Stories

There have been a number of studies that have tried to establish that some people are hypersensitive to EMFs. Ben Goldacre of Bad Science writes:

> There have been 31 studies looking at whether people who report being hypersensitive to electromagnetic fields can detect their presence, or whether their symptoms are worsened by them. A typical experiment would involve a mobile phone hidden in a bag, for example, with each subject reporting their symptoms, not knowing if the phone was on or off.
>
> Thirty-one is a good number of studies, and 24 found that electromagnetic fields have no effect on the subjects. But seven did find a measurable effect...in two of those studies with positive findings, even the original authors have been unable to replicate the results; for the next three, the results seem to be statistical artifacts; and for the final two, the positive results are mutually inconsistent (one shows improved mood with provocation, and the other shows worsened mood).
> <tinyurl.com/3jyx4s8>

At this time, it looks as if electrosensitivity is a psychosomatic disorder. For example, a research team in Norway (2007) conducted tests using sixty-five pairs of sham and mobile phone radio frequency (RF) exposures. "The increase in pain or discomfort in RF sessions was 10.1 and in sham sessions 12.6 (P = 0.30). Changes in heart rate or blood pressure were not related to the type of exposure (P: 0.30–0.88). The study gave no evidence that RF fields from mobile phones may cause head pain or discomfort or influence physiological variables. The most likely reason for the symptoms is a nocebo effect."

Robert Pool claims popular opinion has been aroused against EMFs by unscientific sources such as *The New Yorker* magazine (Pool 1990). Paul Brodeur, called "a scientifically-ignorant writer" by physicist Bob Park, wrote three fear mongering and scientifically inept articles for *The New Yorker* in the early 1990s and published them as a book in 1993. The fear that cell phones might be causing brain tumors was aroused by ABC's "20/20" (October 1999) in a story that focused on the claims of Dr. George Carlo, who for the previous six years ran the cell phone industry's research program on the effects of radiation from cell phones. Gordon Bass also relied heavily on Carlo for his alarmist piece in *PC Computing*, "Is Your Cell Phone Killing You?" (November 30, 1999). Carlo contradicts the conclusions of most other researchers in the field and maintains that "we now have some direct evidence of possible harm from cellular phones." Carlo also claims there is a causal connection between Wi-Fi and autism.

In a press release on October 20, 1999, the FCC responded to "20/20" and claimed that the "values of exposure reported by ABC were well within [the] safety margin, and, therefore, there is no indication of any immediate threat to human health from these phones." Furthermore, the "20/20" story claimed that cell phone antennae emit radiation into the brain, which is misleading. You might also say that TVs and radios emit radiation into the brain if you put your head close enough to those devices.

Talk show host Larry King also aroused fear of a cell phone/brain cancer connection. King introduced the nation to a widower who claimed that his wife's fatal brain tumor was caused by the EMF emitted from her cellular phone, even though she had been using the phone for only six months. There was a lawsuit, of course. The evidence? The tumor was located near where she held the phone to her ear. The major networks reported the story about the lawsuit, the brain tumor, and the cellular phone. Scientists were interviewed to give the story more 'depth' and credibility. However, no scientist has yet found a causal connection between EMF and cancer, much less between cellular phones and brain tumors. A scientist who has exposed existing tumors to EMF was interviewed. He reported that his research indicates that tumors grow faster when exposed to EMF. Sales of cellular phones dropped and stock in companies that manufacture them dropped. Because tumors exposed to EMF grow more rapidly than tumors not so exposed does not indicate that EMF causes tumors, cancerous or otherwise.

Pool also reports that "there have been numerous scientific reports of elevated levels of leukemia in people who are exposed to high EMF levels on the job, such as power-line repairmen and workers in aluminum smelters." Because there have been and probably will continue to be a few studies that find some sort of correlation between cancer and living by power lines, lawyers will always be able to find some evidence, however weak, to justify filing a lawsuit. Over 201 challenges to utility projects were made in 1992 in which EMF was an issue. At least three suits have been filed in federal courts claiming exposure to utility lines caused cancer (Pool, 1991). Utility companies were scared. They poured billions of dollars into efforts to cut EMF exposure from their power lines. Dr. Robert Adair, a physicist at Yale University, calls the reaction "electrophobia" and says that it would take EMF levels 150 times higher than those measured by the Swedish researchers to pose a hazard.

Lawyers can take their cases to court long before the scientific evidence is anywhere near conclusive. And the standards of proof in a court of law are appallingly much lower than those in science. "All it's going to take is one or two good hits and the sharks will start circling," says Tom Ward, a Baltimore attorney who sued Northeast Utilities Co. and its Connecticut Light & Power Co. unit over an alleged EMF cancer (Pool, 1991). The panic led to a great push to bury all power lines. Better safe than sorry? The cost goes up twenty-fold to bury the lines. Then what? Lawyers claiming their clients' cancers were caused by EMFed water? It is bad enough trying to sell a house with power lines nearby when people care about the ugliness of the view. But try to sell the same house when people are afraid of getting cancer from the ugly lines! In any case, we will have to bury our electrical wires even deeper than our power poles are high if we are to make a significant difference in shielding us from the magnetic fields of power lines.

Conclusion

It is not very likely that the average person has anything to worry about from power lines, cell phones, microwave ovens, cordless phones, baby monitors, or Wi-Fi. Most of us do not get that close to power lines to be significantly affected by their EMFs. Our exposure to them, even if they are nearby, is not direct, up close, and constant. The energy emitted by cell phones, cordless phones, and baby monitors (10 milliwatts) is pretty weak. There is more EMF exposure from the wiring in our homes and the electrical appliances we use than from our cell phones or Wi-Fi. No one can avoid electromagnetic radiation. It is everywhere. We are constantly exposed to it from light, commercial radio and television transmissions, police 2-way transmissions, walkie-talkies, etc. Furthermore, "while electrical fields are easily screened, magnetic fields make their way unimpeded through most substances" (Pool, 1990). In fact, it is curious that while fear of EMFs is on the rise so is magnet therapy as a panacea and source of positive energy for the healthful-minded New Ager.

SOURCES FOR APPENDIX A 277

Appendix B
UFOs and Interstellar Space Travel

It is probable that there is life elsewhere in the universe and that some of that life is intelligent. There is a high mathematical probability that among the gazillions of stars in the billions of galaxies there are millions of planets in age and proximity to a star analogous to our Sun. The chances seem very good that on some of those planets life has evolved. We should not forget, however, that the closest star (besides our Sun) is so far away from Earth that travel between the two would take more than a human lifetime. The fact that it takes our Sun about 200 million years to revolve once around the Milky Way gives one a glimpse of the perspective we have to take of interstellar travel. We are 500 light-seconds from the Sun. The next nearest star to Earth's Sun (Alpha Centauri) is about 4 light-years away. That might sound close, but it is actually something like 24 trillion miles away. You might think we could get there in just four years if we could travel at the speed of light. Traveling at the speed of light is easy for a photon, but our fastest spacecraft, Voyager, travels at about 38,500 mph (62,000 km/h). It would take about 70,000 years to get to Alpha Centauri on Voyager.

Even traveling at one million mph, it would take more than 2,500 years to get there. (Imagine the ancient Greek philosopher and mathematician Pythagoras getting on a spaceship in 500 BCE. Some 50 generations later one of his descendants is born on the ship as it pulls into orbit around a star in Alpha Centauri in 2011.) To get there in twenty-five years would require traveling at more than 100 million miles an hour for the entire trip. If you could travel at 100 million miles an hour you could make it to the Sun and back in about an hour and 45 minutes. The spacecraft would have to be built of some mighty fine stuff to endure such speeds for such a length of time. And there'd be no repair shops on the way. But the main problem for travel between stars is the fuel, the energy needed to get there. Estimates of the energy needed to travel to a star 4 light-years away range from one to one hundred times the total energy output of our entire planet for a year. We couldn't do it with nuclear fission because our spacecraft would have to carry thousands or millions of nuclear power plants on its wings! We could do it if we could produce energy the way the Sun does, by nuclear fusion. If we could collect and use all the energy the Sun produces in just one second, we'd have enough energy to last about 500,000 years! Unfortunately, we'd burn to a crisp if our spaceship produced energy the way the Sun does. Maybe someday we'll figure out a way to provide enough energy to travel to a planet outside of our solar system. And maybe some intelligent beings out there in deep space have figured it out, but I wouldn't bet my calipers on it.

When you consider that entire galaxies can pass through one another without any stars colliding, you can get some idea of how empty the universe is. If you got aboard a rocket ship equipped with an endless supply of energy and shields that deflected all harmful rays as well as the gravitational pull of stars and planets, you might travel forever and never come close enough to anything even remotely habitable.

Despite the probability of life on other planets and the possibility that some of that life may be very intelligent, any signal from any planet in the universe broadcast in any direction is unlikely to be in the path of another inhabited planet. It would be folly to explore space for intelligent life without knowing exactly where to go. Yet, waiting for a signal might require a wait longer than life on any planet might endure. If you were a magical massless being who hopped aboard a photon shot from Earth in any direction, it is highly likely that you would pass through the entire universe without hitting anything larger. Also, if we do get a signal, the waves carrying that signal left hundreds or thousands of years earlier and by the time we tracked its source down, the sending planet may no longer be habitable or even exist.

Thus, while it is probable that there is intelligent life in the universe, traveling between solar systems in search of that life poses some serious obstacles. Such travelers would be gone for a very long time. We would need to keep people alive for hundreds or thousands of years. We would need equipment that can last for hundreds or thousands of years and be repaired or replaced in the depths of space. These are not impossible conditions, but they seem to be significant enough barriers to make interstellar and intergalactic space travel highly improbable. The one thing necessary for such travel that would not be difficult to provide would be people willing to make the trip. It would not be difficult to find many people who believe they could be put to sleep for a few hundred or thousand years and be awakened to look for life on some strange planet. They might even believe they could then gather information to bring back to Earth where they would be greeted with a ticker tape parade down the streets of whatever is left of New York City.

Appendix C
Acupuncture, CAM (complementary and alternative medicine), and Faith

In my book *The Skeptic's Dictionary* and on the website of the same name (www.skepdic.com) I made several erroneous claims about acupuncture and other alternatives to scientific medicine. I would like to take this opportunity to correct those errors.

I was way wrong about acupuncture. First, I was wrong to claim that acupuncture has been practiced in China for more than 4,000 years. The earliest manuscripts of Chinese medicine date from the second century BCE and they make no mention of acupuncture. A tomb of a Chinese prince dating from the second century BCE contained a set of four gold and five silver needles, but it is speculation that the needles were designed for acupuncture. Stone needles thought to be 5,000 years old have been found in a tomb in Mongolia, but we don't know how the needles were used. Ancient cultures around the world have used needles for such things as tattooing, scarifying, burning, cauterizing, lancing, piercing, and bloodletting. Did the Chinese teach the rest of the world these things? Did they learn them from others? Did they develop independently in Europe, Egypt, Arabia, and other places? When and where acupuncture began is unknown. The technology for needles made of spun steel, which today's acupuncture needles are usually made of, didn't exist until the early 17th century. In any case, the word 'acupuncture' is clearly not Chinese, but Latin. Acus means "needle" and pungere means "to prick." The first use of the term in the West was in the late 17th century, but the first use that also connected needling with chi, meridians, yin and yang, was by the 20th century Frenchman George Soulié de Morant. (This history of acupuncture is based on Imrie, n.d.)

Morant spent nearly twenty years in China at the beginning of the twentieth century. He spent the next forty years actively promoting acupuncture among medical professionals in Europe. Just before his death in 1955, he completed *L'Acuponcture Chinoise*, which introduced the notions of qi (chi) as energy (or life force) and meridians as the pathways of qi. In 1943, the first society of acupuncturists in the West was founded in Paris. Auricular acupuncture was invented by French physician Dr. Paul Nogier in the 1950s. Nogier saw the ear as an inverted fetus and postulated that the ear is a map of corresponding bodily organs.

While acupuncture was being promoted in the West as an ancient healing art that could cure just about anything, it was being banned in China and Japan. After the introduction of scientific medicine in those countries, efforts were made to stifle ancient medical superstitions and myths. By 1911 in China acupuncture was no longer a subject for examination by the Chinese Imperial Medical Academy. Mao Zedong promoted Chinese traditional medicine for political and practical reasons, but he did not use it or believe in it himself. Acupuncture came to the attention of the Western world in dramatic fashion when it was widely reported in 1971 that *New York Times* journalist James Reston had undergone an appendectomy in Beijing with the only anesthesia being provided by acupuncture. In fact, he had chemical anesthesia for the operation and acupuncture was administered *afterward* to relieve pain. Reston allegedly reported that about an hour after the acupuncture treatment he felt pain relief. Was the relief due to the acupuncture? Perhaps. It may also have been due to his having a bowel movement. Did the acupuncture cause his bowel movement? I don't know, but I do know that after this story was reported in the Western press, acupuncture began its current run as the darling of alternative medicine in the West. Simultaneously, acupuncture has grown less popular in China. It might be of interest to some readers that The National Council Against Health Fraud (NCAHF) found that of the forty-six medical journals published by the Chinese Medical Association, not one is devoted to acupuncture or other traditional Chinese medical practices.

Perhaps I was wrong to define acupuncture as "a traditional Chinese medical technique for unblocking chi by inserting needles at particular points on the body to balance the opposing forces of yin and yang." While many proponents of acupuncture consider this description to be accurate, many others think it is misleading to theorize about *how* acupuncture works. Perhaps, they say, it works by blocking pain signals or by releasing endorphins or by some other unknown physical mechanism. To bring in chi flowing along meridians and balancing yin and yang is to confuse the issue, according to some people.

I was certainly wrong to have stated that chi allegedly flows "through the body along fourteen main pathways called meridians." Tradition has it that there are twelve main pathways and some minor pathways. I still do not see any good reason for believing that chi is an energy that allegedly permeates all things. There are a number of energy therapies that make this claim. Oddly, they produce similar results in practice and in laboratory tests. Scientific studies over the past few years have supported my original position: the beneficial effects of acupuncture and other forms of energy medicine are probably due to "a combination of expectation, suggestion, counter-irritation, operant conditioning, and other psychological mechanisms." Similar results have been obtained for true acupuncture, sham acupuncture (pretending to stick needles into a person), acupressure (where acupoints are touched but not needled), reiki and therapeutic touch (where the therapist allegedly manipulates chi without touching the patient at all), Tong Ren (a kind of voodoo acupuncture where one strikes acupoint marks on a doll with a hammer to release energy), and distant healing (where the healer doesn't need to be in the physical presence of the patient). Apparently, as long as the patients believe they are getting energy treatment, they get some relief, but it doesn't really matter whether the patient is stuck with needles, touched or not, or even in the presence of the healer. Of course, there is no way to disprove the claims that coming near the acupoints, thinking about them, or hitting effigies of them triggers the unblocking of chi. But such explanations seem superfluous when there are simpler explanations that can plausibly account for the same data: placebo effects and non-specific effects. In fact, developments in modern physics and biology since the 19th century have rendered unnecessary all forms of vitalism and explanations of biological processes in terms of energies that can't be measured by any scientific instrument but can be felt by something much less sensitive: the human hand.

The evidence from many high caliber scientific studies has shown that many forms of energy healing relieve many people of many symptoms and that this is probably due to one or more of the following factors, some of which are referred to as placebo factors, some as *false* placebo factors (because in some studies their effects have been erroneously attributed to the placebo effect):

- classical conditioning
- suggestion by the healer
- patient beliefs in the competence of the healer and in the method of healing
- patient expectancy and hope for recovery
- the healer's manner (showing attention, care, affection, sincerity, knowledge)
- the color of the treatment room or the color of the pill one is given (might affect patient expectancy)
- the rituals and theater involved in the delivery of the treatment, including technical jargon, special uniforms, medical gadgetry, treatment room set-up, and the like
- spontaneous improvement (the pain or illness runs its natural course to its natural conclusion)
- fluctuation of symptoms
- regression to the mean
- additional simultaneous treatment from scientific medicine
- patient politeness or subordination (the patient doesn't want to disappoint the healer)
- neurotic or psychotic misjudgment
- psychosomatic phenomena

It is possible, of course, that energy healers are affecting the balance of yin and yang by the butterfly effect or some other magical procedure, but such explanations seem unnecessary and farfetched.

Another error I made was in referring to many alternative therapies as "useless" or "ineffective." Most of these therapies that now go under the heading of CAM, complementary and alternative medicine, are useful and effective. However, they are no more useful or effective than placebos or doing nothing. In fact, the expression "alternative and complementary" seems designed to describe treatments that seem to have a positive effect but have not been shown to have any effect beyond a placebo effect or no intervention. There seems to be no harm in offering them *in addition to* treatments based on scientific medicine. Homeopathic remedies, for example, may be inert and have nothing in them but water or alcohol, but because of conditioning and other placebo factors, even inert substances can have a positive (or negative) effect on people. Some studies have found that acupuncture and the placebo treatment of sham acupuncture can stimulate an opioid response and lead to the release of natural painkillers. Many people do not understand that placebos can have physical effects. They think that if a placebo has an effect, it must be "all in their head." Not true. The effect can be in their arm or foot or left ear.

On the other hand, I was correct in claiming that "Traditional Chinese medicine is not based on knowledge of modern physiology, biochemistry, nutrition, anatomy, or any of the known mechanisms of healing." What has happened is that many current proponents of acupuncture try to retrofit traditional acupuncture to modern scientific knowledge. This procedure is little more than a game of confirmation bias: looking for ways to confirm what is already believed. Most people do not realize how easy it is to make up an explanation that fits with one's beliefs. To do so is rather trivial, however, from a scientific point of view, where the goal should be to try to *falsify* claims rather than confirm them. It is possible that all ancient cultures had advanced medical knowledge but lost it due to natural or human disasters. It seems more plausible, however, that our ancestors were scientifically illiterate, as is indicated by such misconceptions as that the heart functions as the seat of consciousness and memory. To speculate that the ancients knew more than modern scientists isn't justified. Our ancestors knew a lot of things about plants as medicines—knowledge they acquired by trial and error. We have better methods today and our results, while imperfect, are much more reliable than those of the village shaman or witch doctor two millennia ago. In fact, it is because of our understanding of the complexity of placebo effects, conditioning, and the other factors listed above, that we are able to grasp why the village shaman could be successful, even though his medicine bag was little more than a bag of tricks. Most illnesses don't kill you. You will recover from almost anything that ails you whether you get treatment or not. A little song and dance, a puff of smoke here or there, some magic tricks on occasion, accompanied by appropriate dress, rituals, and incantations and *voila*: the cure. I was wrong in suggesting, however, that the patients of practitioners of scientific medicine don't benefit from placebo and false placebo factors. All healing—scientific, pseudoscientific, and magical—benefits from placebo and false placebo effects.

While I was wrong to suggest that CAM therapies are useless, I was not wrong to claim that "integrative medicine" is "quackery mixed with scientific medicine." The term "integrative medicine" was popularized by Andrew Weil, M.D., a graduate of Harvard Medical School but one who did not complete a residency nor, as far as I can ascertain, ever take the U.S. medical boards. (He did a one-year internship at Mt. Zion Hospital in San Francisco.) Instead of practicing medicine, Weil traveled and studied indigenous medicine as practiced in South Dakota and various places in South America. For the past thirty years, he has made his living writing and lecturing on alternative medicine. Dr. Weil founded the Arizona Center for Integrative Medicine at the University of Arizona. Weil integrates scientific medicine with Ayurvedic medicine (herbal and dietary medicine allegedly originating in ancient India) and other "natural" cures. One of his main tenets is: "It is better to use natural, inexpensive, low-tech and less invasive interventions whenever possible." There is no compelling scientific evidence for the claim that natural interventions are generally superior to artificial ones. If a natural herb and a powerful pharmaceutical have the same medicinal effect, the herb will probably have fewer adverse side-effects. But, as far as I know, there are no herbs that have the same medicinal effect as powerful pharmaceuticals. Millions of people use herbs and natural products, such as calcium, echinacea, ginseng, ginkgo biloba, glucosamine, saw palmetto, shark cartilage, and St. John's wort. All of these, when tested scientifically, have failed to support the traditional wisdom regarding their healing powers. Pharmaceuticals and other treatments are much superior to most herbal remedies. If a plant has been shown to be effective as a healing agent, the active ingredient has been extracted and tested scientifically and is part of scientific medicine. Otherwise, any beneficial effect following use of the herb or plant is probably best explained as due to the placebo effect, natural regression, the body's own natural healing processes, or to some other non-herbal factor. It should be noted that many people take vitamin and mineral supplements, both natural and synthetic, in the mistaken belief that there is sound scientific evidence that such supplements contribute to well-being or can prevent cancer or heart disease. So far, the scientific evidence overwhelmingly indicates that vitamin and mineral supplements provide no health benefits in general. In fact, some supplements have been linked to negative consequences.

The appeal of Weil's integrative medicine is that he mixes sound scientific knowledge and advice with illogical hearsay. For example, when I checked his Men's Health Internet page in 2008, he provided scientific information regarding men with prostate problems. He offered common-sense advice such as don't ingest caffeine and alcohol if you are having trouble with frequent urination, since these substances will increase the need to urinate. But he also advised men to eat more soy because: "Asian men have a lower risk of BPH and some researchers believe it is related to their intake of soy foods." BHP stands for *benign prostatic hyperplasia* (an enlargement of the prostate that occurs with aging.) He ignored the scientific evidence that there are high rates of cancers of the esophagus, stomach, thyroid, pancreas and liver in Asian countries. Should we blame these high rates on the consumption of soy? Of course not. Correlation is not causation. Weil also states that saw palmetto "may help" BPH because: "There is clinical evidence that saw palmetto can help shrink the size of the prostate, and it may help promote healthy prostate function." There is also strong clinical evidence that saw palmetto *doesn't* help shrink the size of the prostrate. It is not good medicine to pick and choose only those studies that support your biases.

Another thing I was wrong about was in underestimating the power of experience to deceive us by a variety of cognitive and perceptual illusions to the point where we refuse to accept the evidence from scientific studies if that evidence doesn't support a strongly held belief. I was also wrong, however, to put too much trust in scientific studies. I was right to suggest that single studies in medicine rarely justify drawing grand conclusions about anything positive, but I should have encouraged more skepticism regarding scientific studies in general.

Infertility, Meta-analysis, and the Media

It's clear from the testimonial and scientific evidence that acupuncture benefits some people some of the time for some conditions, particularly for the relief of pain. It's also clear that acupuncture doesn't benefit anyone for some conditions, even though there are published studies that conclude otherwise. The evidence tells me that it is criminal, for example, to treat infertility with acupuncture. Bill Reddy, a practicing acupuncturist, disagrees. Reddy believes that "acupuncture is a thoroughly proven system of healthcare," as evidenced by the fact that PubMed alone lists some 13,000 published studies on acupuncture and that some of them support his belief that acupuncture is an effective treatment for infertility. Reddy claims that "countless studies have proven acupuncture's effectiveness in improving the viability and diameter of ova." He selects one such study for discussion. I assume he selected it because it is typical or he thinks it is one of the better studies. It was published in 1993 in the *Journal of Chinese Medicine* by Mo et al. Reddy notes that the "total effective rate was 82.35%," whatever that might mean. He quotes from the article, but he seems to gloss over the fact that the researchers are very cautious in their claims, using the word 'may' to qualify their conclusions:

...the results also showed that acupuncture *may* adjust FSH, LH, and E2 in two directions and raise the progesterone level, bringing them to normal. The animal experiments confirmed this result. Results showed that acupuncture *may* adjust endocrine function of the generative and physiologic axis of women, thus stimulating ovulation. (Emphasis added.)

Furthermore, this study had no control group and was small (34 patients). The authors also make some unsubstantiated claims that Reddy doesn't mention, e.g., that acupuncture at the Chong and Ren channels "nourishes uterus to adjust the patient's axis function and recover ovulation." Also, we should note that researchers at the University of Oklahoma studied more than ninety-seven patients who were getting in-vitro fertilization, some of whom were also getting acupuncture twenty-five minutes before and after the embryo was transferred from the test tube to the womb. The pregnancy rate of the group that did *not* receive acupuncture was 69.9 percent, while the pregnancy rate of the group that did receive acupuncture was 56.2 percent. These data strongly indicate that acupuncture has no positive effect on fertility.

Defenders of the view that acupuncture assists IVF cite a meta-study to support their position, but there is also a meta-study that does *not* support their position. It's easy to see why there would be conflicting meta-studies. A meta-analysis is a study that lumps together the data from several independent studies and does a statistical analysis on the data as if they were collected in a single, large study. One of the major problems with meta-studies is that researchers must be selective in choosing which studies to include in their analysis. Some studies will have to be rejected because they are fatally flawed: they're too small, use no controls, didn't randomize the assignment of subjects, or the like. Different researchers will include and exclude different studies. Even if they agree on the criteria used to determine which studies to include, they will often disagree on the application of the criteria. In the end, one will often find two meta-studies that contradict each other and each side will claim the other excluded studies that should have been included or included studies that should have been excluded. A common accusation is that if the researcher got a positive result it was because he excluded too many studies that got negative results. Or, if the researcher got a negative result, it was because he included too many negative studies or didn't include enough positive studies. Furthermore, the media often have no clue as to how to properly evaluate a meta-analysis.

Anyway, a meta-study by Eric Manheimer et al. appeared in the *British Medical Journal* called "Effects of acupuncture on rates of pregnancy and live birth among women undergoing in vitro fertilisation [IVF]: systematic review and meta-analysis." The news media hailed the study as finding evidence that acupuncture improves the chances of successful fertilization. The authors of the study, however, note that the connection between acupuncture and fertilization "is far from proven." They call their evidence "preliminary" and state that it "suggests that acupuncture given with embryo transfer improves rates of pregnancy and live birth among women undergoing in vitro fertilisation." The media erroneously reported that the data showed a 65% increase in fertility in those treated with acupuncture, when the actual figure was closer to 10%. Furthermore, acupuncture researcher Peter Braude claims that "the BMJ paper didn't include all the studies, and if you include the negative ones there is no effect." Braude supervised a team of researchers who recently finished a meta-study on acupuncture and IVF treatment that found no effect. They presented their conclusions in 2008 to the European Society of Human Reproduction and Embryology conference in Barcelona, Spain. The researchers identified eighty-three trials in the medical literature, of which thirteen were found to be of suitable quality to be included in the meta-analysis. The way to avoid these kinds of conflicting reports is to avoid meta-analysis and do single studies that use large samples.

Battlefield Acupuncture

Recently, another application, called "battlefield acupuncture," has been promoted by Dr. Richard Niemtzow, the first official acupuncturist in the U.S. armed forces. Niemtzow believes that inserting tiny semi-permanent needles into very specific acupoints in the skin on the ear blocks pain signals from other parts of the body and prevents them from reaching the brain. He has done some studies but none of them used proper methods: randomized, placebo-controlled studies with at least 50 soldiers as subjects and designed to demonstrate this alleged blocking of pain signals. The United States Air Force taught "battlefield acupuncture" to physicians who were deployed to Iraq and Afghanistan in early 2009. Niemtzow believes ear acupuncture can be used on the battlefield to relieve pain without worry about side effects (including addiction) or adverse reactions, which are apparently a problem with pain-killing drugs.

Niemtzow's enthusiasm for acupuncture far exceeds his evidence for its efficacy. In my opinion, the Air Force is using American soldiers as guinea pigs and will not be providing our men and women on the battlefield with the best that medicine has to offer should they be wounded in battle. Niemtzow's website says he specializes in acupuncture for dry mouth and dry eye, but he also treats obesity and a few other things. He's written that he and his cohorts use at least ten different kinds of acupuncture treatments for the following list of disorders:

> fibromyalgia, protruding disks, reflex sympathetic dystrophy, degenerative disk disease, spinal stenosis, frozen shoulder, peripheral neuropathy secondary to diabetes or chemotherapy, torticollis, overuse syndromes, abdominal pain of unknown etiology, tendonitis, carpal tunnel syndrome, arthritis, osteoarthritis, migraines ... obesity, nicotine abuse, dry mouth and dry eyes from various etiologies, hot flashes, chronic fatigue along with depression, and dermatological conditions such as eczema....

Dr. Niemtzow speculates that acupuncture is a science that may "eclipse Newtonian physics," but this belief is based on little more than his faith in ancient Chinese metaphysics (especially the concept of chi) and a lack of curiosity that leads him to forgo randomized, double-blind, controlled studies to ferret out such things as placebo effects from false placebo effects. His claim that sticking needles in the ear blocks pain signals from battlefield wounds is not based on good science. There are millions of people in pain who would love to be able to block pain signals by putting little pins in their ears. If he could prove it's true, he'd have his Nobel and my thanks as well.

Proper Testing of Acupuncture

What has become increasingly clear from acupuncture studies that use proper controls is that acupuncture is a placebo therapy: the effects of acupuncture are not significantly different from the effects of placebos or of making no intervention. A study that compares one group given acupuncture and another group given pills, exercise, massage, or some other sort of therapy is not a properly controlled study because such a design cannot measure the placebo effect or the effect of doing nothing. A properly controlled acupuncture study is one that is double-blind, randomized, and uses a control group. The only proper control group for an acupuncture study is a group that receives sham acupuncture. A double-blind acupuncture study would be one where neither the therapist nor the patient knows who is getting true acupuncture. (How could the therapist not know who is getting the true acupuncture?) True acupuncture is acupuncture that sticks needles into traditional acupuncture points on the body to traditional depths. Sham acupuncture is of at least three types. One type inserts needles into non-traditional sites at the same depth as traditional acupuncture. Another does the same but to a shallower depth. The third type uses a method that prevents the needles from actually being inserted into the body. It is important that the patients in the sham group think they are getting true acupuncture. It is equally important that the acupuncturist not indicate to the subjects in any way whether she is delivering true or sham acupuncture. Any study where the patients can easily detect whether they are getting true acupuncture is an invalid study because it cannot measure the placebo effect.

Some researchers obviously do not understand the placebo effect. When they compare true and sham acupuncture groups to a third group getting some other kind of therapy and find that the two acupuncture groups show a significant positive effect over the third group, they don't realize such a result supports the position that acupuncture works by the placebo effect.

A proper acupuncture study would randomly assign at least 50 patients to either a true or sham acupuncture group and not reveal to the patients which group they have been assigned to. Comparing an acupuncture group to a group that does not get acupuncture is an invalid study, unless one is trying to measure the different degrees of effectiveness of placebo treatments. A proper acupuncture study should not have a high dropout rate and would, if appropriate, involve follow-up reviews to measure long-term effectiveness.

Homeopathy

I was wrong about homeopathy on at least two counts. One, I should have been more vigorous in criticizing the work of Jacques Benveniste and the claim that homeopathic remedies, which are nothing but water, work because water has a "memory" of some active ingredient that it once had contact with. And I should have been more emphatic in claiming that homeopathy's long list of satisfied customers is probably due to a set of beliefs and rituals that get lumped under the vague heading of "the placebo effect."

Scientists like Jacques Benveniste (1935-2004), who claim to know how homeopathy works, have put the cart before the horse. Benveniste claimed to have proven that homeopathic remedies work by altering the structure of water, thereby allowing the water to retain a "memory" of the structure of the homeopathic substance that had been diluted out of existence. The work in Benveniste's lab was thoroughly discredited by a team of investigators sent by *Nature* to evaluate an attempted replication of Benveniste's work. Neither Benveniste nor any other advocate of the memory-of-water speculation has explained how water has forgotten all the other billions of substances its molecules have been in contact with over the millennia, but it remembers just the contact with the homeopathic substance. (Maybe homeopaths have discovered a magical method of "degaussing" memories from water.) Benveniste claimed that a homeopathic solution's biological activity can be digitally recorded, stored on a hard drive, sent over the Internet, and transferred to water at the receiving end. He was a successful biologist working in a state-run lab until he started making such claims, which cost him his status and reputation as a reputable scientist. Since homeopathic remedies are inert, there is no need for a physical or mechanistic explanation as to how they work. What there is need of is an explanation for why so many people are satisfied with their homeopath despite all the evidence that homeopathic remedies are inert and no more effective than a placebo.

There have been several reviews of various studies of the effectiveness of homeopathic treatments and not one of these reviews concludes that there is good evidence for any homeopathic remedy (HR) being more effective than a placebo. Homeopaths have had over two hundred years to demonstrate their wares and have failed to do so. There are single studies that have found statistically significant differences between control groups and groups treated with an HR, but none of these have been replicated or they have been marred by methodological faults.

An evaluation of five reviews of homeopathic studies has been done by Terence Hines who found that more than one hundred studies have failed to come to any definitive positive conclusions about homeopathic potions. There have been at least twelve scientific reviews of homeopathy published since the mid-1980s. Guess what? Homeopathic remedies are not more effective than placebos or no intervention (Ramey 2000). Why is homeopathy so popular, then? One reason is the prevalence of a misunderstanding of the causes of disease and how the human body deals with disease. Samuel Hahnemann, the father of homeopathy, was able to attract followers because he appeared to be a healer compared to those who were cutting veins to bleed out bad humors or using poisonous purgatives to balance humors. More of his patients may have survived and recovered not because he healed them, but because he didn't infect them or kill them by draining needed blood or weaken them with strong poisons. Hahnemann's medicines were essentially nothing more than common liquids and were unlikely to cause harm in themselves. He didn't have to have too many patients survive and get better to look impressive compared to his competitors. If there is any positive effect on health it is not due to the homeopathic remedy, which is inert, but to the body's own natural curative mechanisms, the beliefs of the patient, and the manner of the homeopath.

Stress can enhance and even cause illness. If a practitioner has a calming effect on the patient, that alone can result in a significant change in the feeling of well-being of the patient. That feeling might well translate into beneficial physiological effects. The homeopathic method involves spending a lot of time with each patient to get a complete list of symptoms. It's possible this attentiveness has a significant calming effect on some patients. This effect could reduce stress and enhance the body's own healing mechanisms in some cases. As homeopath Anthony Campbell (2008) puts it: "A homeopathic consultation affords the patient an opportunity to talk at length about her or his problems to an attentive and sympathetic listener in a structured environment, and this in itself is therapeutic." In other words, homeopathy is a form of psychotherapy:

> ....most homeopaths like to allow at least 45 minutes for a first consultation and many prefer an hour or more. Second, patients feel that they are being treated "as an individual". They are asked a lot of questions about their lives and their likes and dislikes in food, weather, and so on, much of which has no obvious connection with the problem that has led to the consultation. Then the homeopath will quite probably refer to an impressively large and imposing source of information to help with choosing the right "remedy". (Campbell)

We know that the sum of all the scientific evidence shows clearly that homeopathic remedies are no more effective than placebos. This does not mean that patients don't feel better or actually get better after seeing a homeopath. That is quite another matter and is clearly the reason for the many satisfied customers.

So, I was wrong to imply that most homeopathic "cures" are mostly due to misdiagnosis, spontaneous remission (the disease running its natural course), natural regression, or some other therapy being used along with homeopathy. The bulk of the satisfied customers are probably satisfied because they believe the homeopath knows what he's doing and that the medicine is effective. The patient believes in the homeopath because he appears to have a vast knowledge of remedies, has impressive books and shelves of potions for everything under the sun, is calm and confident, has many success stories to share, gives hope and confidence to the patient, relaxes and eases the stress the patient may be experiencing, and a number of other things that generally are lumped together as "the placebo effect." It is easy to understand why a homeopathic physician, with hundreds or thousands of satisfied customers, would read the scientific literature differently than an independent, unbiased observer who doesn't care one way or the other whether homeopathy is superior to placebo medicine. The homeopath with years of experience seeing his patients helped by his medicine will be highly motivated to latch onto studies that find homeopathy works and to be less critical of studies that find the opposite. It is easy to rationalize the methodological faults of studies that don't support one's hypothesis, and it is easy to gloss over the faults of studies that fit with one's beliefs. To settle the issue fairly, however, it is necessary that a large-scale, randomized control study be done that is designed to measure placebo effects and demonstrate that at least some of the effects from the homeopathic remedy are not due to placebo effects.

Placebo Medicine vs. Scientific Medicine

To those acupuncturists and homeopaths who come to realize that their medicine works, but is a placebo treatment, you have a decision to make. You can act as one shaman did when he realized his medicine worked no matter what he did: you can continue with the rituals and arcane ceremonies associated with your art. It would be easy to rationalize, since you are helping people. You may even be helping people who otherwise wouldn't get any treatment from anyone. You'll have many satisfied customers and may make a decent living as well. You'll get a lot of communal reinforcement from other practitioners, the popular media, journalists, and celebrities.

The danger from acupuncture is that it is being promoted as superior to scientific medicine, when in fact it is clearly inferior. Acupuncture is touted as appropriate for almost any disorder or disease in man or beast, when the evidence clearly shows that such a belief is a dangerous delusion. People go to acupuncturists for treatment of AIDS, allergies, arthritis, asthma, Bell's palsy, bladder and kidney problems, breast enlargement, bronchitis, colds, constipation, cosmetics, depression, diarrhea, dizziness, drug addiction (cocaine, heroin), epilepsy, fatigue, fertility problems, fibromyalgia, flu, gynecologic disorders, headaches, high blood pressure, hot flushes, irritable bowel syndrome, mental illness, migraines, nausea, nocturnal enuresis (bedwetting), pain, paralysis, post-traumatic stress disorder (including rape victims), PMS, sciatica, sexual dysfunction, sinus problems, smoking, stress, stroke, tendonitis, vision problems, and just about anything else that might ail a human being. We should be skeptical of any modality that claims such versatility. Not only can it be used for hundreds of different kinds of ailments, acupuncture as practiced in China is not the same as that practiced in Korea or Japan or many other places. Even acupuncturists trained in the same tradition will evaluate and treat the same patient differently. R. Barker Bausell, an expert in biostatistics, served for five years as the director of research at the University of Maryland's NIH-funded Complementary Medicine Program (now called the Center for Integrative Medicine). He found it disconcerting that there was no consistency in either diagnosis or treatment recommendations among three "experienced TCM (traditional Chinese medicine) physicians" who examined the same group of patients identified as having rheumatoid arthritis. Bausell knows how to tell a well designed and implemented medical study from a faulty or incompetent one, and he knows how to evaluate the statistical data that make up the backbone of such studies. But the fact that three experts would disagree so fundamentally about diagnosing and treating the same patients, even when they knew that the patients all suffered from the same disease, led him to conclude that even if there is some physiological basis for acupuncture, "it would be worthless." Why? Because highly trained experts "came up with completely different conclusions when examining the same patients."

Scientific medicine isn't perfect, but it doesn't claim to have a treatment for everything, much less a *single* treatment for everything. Furthermore, scientific medicine has numerous examples of treatments that have been shown to be effective independently of conditioning, placebo, or false placebo factors.

Acupuncture has no such examples. As long as acupuncture is limited to such things as treating nausea from chemotherapy, it is a laughable delusion and will probably be effective for some patients. When acupuncture is used *instead of* chemotherapy to treat cancer, it will become a dangerous delusion. The same should be said of other alternative treatments known to be placebos, such as homeopathic remedies. As long as homeopathy is used to treat nausea or insomnia, it is laughable. But when homeopathy is used to prevent malaria, AIDS, or infertility it becomes a dangerous delusion.

Acupuncture, Homeopathy, Energy Healing Work!

You name your alternative treatment and you will find many advocates ready to testify that it works. Yes, placebos really do work! Only a person who does not understand the placebo effect would claim that acupuncture can't be a placebo since it was effective. We know from the studies that have been done on acupuncture, homeopathy, energy healing, and the like that they are no more effective than a placebo or making no intervention. This means that these treatments will have many satisfied customers as long as people are susceptible to suggestion and classical conditioning, believe in the effectiveness of treatment, and have it administered by someone who is seen as a knowledgeable and trustworthy healer.

Some people are convinced that alternative treatments can't be placebo medicine because they work on animals. If acupuncture is just a placebo, how could it help my dog or horse? Ever hear of Pavlov and classical conditioning? You should look it up. It's also possible that it didn't help your dog or horse, but you perceive it that way to confirm your bias.

Finally, I've said nothing so far about hypnotherapy. That is because I agree with R. Barker Bausell that hypnosis and placebos are "so heavily reliant upon the effects of suggestion and belief that it would be hard to imagine how a credible placebo control could ever be devised for a hypnotism study."

Faith

I understand the dangers of scientific medicine. An acupuncturist or homeopath is not as likely as a science-based surgeon, say, to do great damage by malpractice. There are much greater risks in surgery, anesthesia, and pharmaceuticals than there are in ingesting water, sniffing herbs, or having a few needles stuck in one's ear. There is the occasional chiropractor who does great damage but, for the most part, scientific medicine has the potential for causing greater harm than do most alternatives. On the other hand, scientific medicine also does the most good, if by good we mean helping people live longer, healthier lives. Disease would be rampant without scientific medicine. Life expectancy would be much lower. A simple operation in 1998—insertion of a stent in one of my arteries—has kept me alive for more than a decade. Had the same operation been available to my grandfather and father, they might not have died of heart attacks in their early 50s. Had insulin not been available to my mother, it is very unlikely that she would have lived into her 70s. But for most of the things that ail people—the colds and minor aches and pains—the pill or shot the scientifically trained physician provides may be just a placebo and an alternative therapy would have worked just as well. I was wrong to say or imply that alternative therapies are useless or ineffective. They're not, unless you have something seriously wrong with you. In which case, I suggest you see someone trained in scientific medicine who practices scientific medicine. Jacques Benveniste did. He didn't go to a homeopath for his heart problems. He went to a surgeon. Unfortunately, the surgeon couldn't help him and he died shortly after surgery. Scientific medicine is powerful, but it can't keep you alive forever. For that you need faith.

Faith, I realize, can be very powerful, especially when it promises eternal life. But even when faith promises only the same thing that scientific medicine promises—a longer or healthier life—it is often much more powerful than reason and can become an impenetrable shield against science, skepticism, and critical thinking, as the following story will show.

In 2006 Corinna Borden was diagnosed with stage-four Hodgkin's lymphoma, a treatable form of cancer. The one-year survival rate for Hodgkin's lymphoma is about 92% and the five-year survival rate is about 85%. Borden was 29 when diagnosed. Five years later she published a book about her medical treatment, especially her self-medicating with what she calls 'holistic' methods. She drank Epsom salts with grapefruit juice, followed by a glass of olive oil and a glass of apple juice. She tried various forms of energy healing, nutritional supplements, a macrobiotic diet, and acupuncture. She checked into a clinic in Tijuana, Mexico, for treatment with herbs and laetrile. One energy healer told her that anger, resentment, and unhappiness "feed into your nervous system and send your body conflicting messages." It is not even clear what it would mean for emotions to "feed into your nervous system," but apparently that advice did not completely resonate with Borden, for it made it seem that her cancer was somehow her fault. She did not want to be a *patient* and accept all the passiveness that such a term implies. She wanted to be in control of her disease. Who can blame her?

Part of Borden's book describes the conflicts in her marriage caused by her holistic treatments while her husband pursued his work in scientific medicine. Borden's husband, Walter Parker, was finishing his residency at a Michigan hospital when she was diagnosed. His commitment to science-based medicine was clear, but he knew of Borden's penchant for CAM long before they married. Borden, on the other hand, did not reject science-based medicine. She underwent six weeks of chemotherapy. A CT scan, however, showed that the chemo may not have gotten all the cancer cells. There was a "residual hot spot" near the esophagus that could not be biopsied because it was too close to her vena cava, the major vein in the body. Note: the "hot spot" was not biopsied so it is not known that it indicated cancer cells. The article I read from ABC News wasn't clear about what Borden's next treatment was. Apparently, her medical doctors recommended a bone marrow transplant and she started some sort of clinical trial, but her white blood count dropped too low and she didn't finish the trial. In any case, she was in remission at the time her book was published in 2011. Both Borden and her medical doctor husband credit her holistic treatments with her remission. I credit their faith in her beliefs as the main reason the medical treatments and her body's own mechanisms are given short shrift when the credit for her remission is given. Corinna Borden's book is called *I Dreamt of Sausage*. I am glad she is in remission, but she hasn't provided a shred of good evidence that her holistic treatments had anything to do with it. Had she not also received chemotherapy, her book most likely would never have been written.

SOURCES FOR APPENDIX C 278

Appendix D
Critical Thinking

My introduction to the concept of critical thinking came in 1975. I was fresh out of graduate school and teaching my first logic class at Lassen College, a small community college on the eastern slope of the Sierra Nevada. Not only was it my first logic class but it was the first logic class ever taught at Lassen College. Even though I had been hired as a one-year replacement for a teacher who was going on sabbatical leave, I was permitted to submit the course to the curriculum committee. The course that was approved was based on the "baby logic" class I had taken several years earlier at the University of California at San Diego. If I remember correctly, about twenty-five students signed up for the class and three finished. One of the three stayed because he liked me. Another stayed because he didn't know how to drop a class. The remaining student understood the material in the text, *An Introduction to Logic* by Irving Copi. My other courses didn't suffer the same rate of attrition as the logic class, so I concluded—rightly or wrongly—that the fault was not with my teaching but with the course I had designed. Then two things happened that changed my life forever.

I was rehired to replace another teacher who was going on sabbatical and I received a review copy of a new text by Howard Kahane called *Logic and Contemporary Rhetoric: The Use of Reason in Everyday Life*. I read Kahane's book and decided to try it, even though it bore little resemblance to traditional logic texts. The text is now in its 11th edition and a blurb for that edition notes that it "puts critical thinking skills into a context that students will retain and use throughout their lives." A blurb about the author notes that Kahane was one of the founders of the critical thinking movement. That may be true, but nobody called his book a critical thinking text when it was first published. At that time, two other expressions were vying for primacy: 'informal logic' and 'practical logic.' Kahane's text did not concern itself with *forms* of arguments, as did most traditional logic texts. It dealt with real arguments in ordinary English. I should note that Kahane had already published a formal logic text—it, too, is in its 11th edition. The informal logic text contained no mention of Aristotle, no syllogisms, no Venn diagrams, no truth tables, and no sentential or predicate logic. There were no tedious exercises asking the student to symbolize ordinary language arguments and test them for validity. Instead, there were discussions of numerous affective, cognitive, and perceptual biases and illusions. An entire chapter was devoted to how language can be used to manipulate thought. Other chapters focused on non-traditional topics for a logic course: advertising, mass media manipulation, cultural indoctrination through history textbooks, and informal fallacies like those covered here in Chapter Seven.

I have looked through the table of contents for the 11th edition of *Logic and Contemporary Rhetoric* and have seen what can happen to an excellent book when a publisher tries to extend its market. I wouldn't recommend the current edition if I were asked to recommend a text for an introductory course on critical thinking. The most important chapter in the first edition—the one on language—has been moved from Chapter Two to Chapter Seven, and Chapter Two now includes material on syllogisms and deductive validity. Discussion of psychological hindrances to critical thinking has been relegated to Chapter Six. I recall them being introduced very early on in the first edition. In any case, it was Kahane's text that got me started on what we now call *critical thinking,* even though he never used that term.

Kahane introduced a generation of teachers to what would come to be known as critical thinking in the 'strong sense'. He reminded us how one's worldview can be a major hindrance to being fair-minded. A minimum requirement of fair-mindedness is a willingness to take seriously viewpoints opposed to your own. In other words, you have to be willing to admit that you might be wrong. To exempt one's own worldview from critical evaluation is common enough, but if we want to teach our students to think critically we must teach them to try to understand how one's worldview is likely to be embedded with prejudices, biases, and false notions. We have to remind our students that everything we experience or remember is filtered through that set of beliefs and values that make up one's worldview. To think critically is to be willing to examine conflicting positions in a fair-minded way and to accept that even beliefs you've held all your life might be wrong. If you can't do that, you might still be able to develop some critical thinking skills like comparing and contrasting ideas or comparing ideals with practices, but you would be a critical thinker only in the weak sense of being able to apply one or more of the standards of critical thinking in a skillful way. Critical thinking in the strong sense requires that the thinker have a certain *disposition* as well as being knowledgeable of the many affective, cognitive, and perceptual biases that inhibit and distort judgment.

A typical list of critical thinking standards and skills might look like this:

Standards
- Clarity
- Accuracy
- Relevance
- Completeness
- Significance
- Fairness
- Sufficiency of evidence
- Consilience
- Logic: coherence, contradiction, and validity.

Skills
- Abuses of language: doublespeak; understanding vagueness, ambiguity, and obscurity; effective use of definitions
- Recognizing assumptions and implications
- Evaluating sources of information
- Evaluating claims and arguments
- Common fallacies: of assumption, of relevance, of omission, of insufficient evidence
- Evaluating inductive reasoning: simple sampling and analogical reasoning
- Evaluating explanations and causal reasoning
- Evaluating scientific and conceptual theories

- Applying the hypothetico-deductive model and argument to the best explanation.

Critical thinking in the strong sense goes beyond applying standards and skills. It requires knowledge of affective, cognitive, and perceptual biases and how they affect interpretations of experience, testimony, and other evidence. Above all, critical thinking in the strong sense requires a certain *attitude* or disposition that might be summarized as follows:

- *Intellectual humility*: a willingness to admit error, change beliefs when warranted, or suspend judgment;
- *Confidence in reason*: a willingness to go wherever the evidence leads;
- *Intellectual curiosity*: a love of exploring new topics, learning new things, gaining knowledge;
- *Intellectual independence*: willingness to examine honestly and fairly the positions of those you disagree with and willingness to question authority, tradition, and majority opinion.

I was first introduced to the notion of critical thinking in the strong sense by Richard Paul, a philosopher at Sonoma State University and now the director of the Center for Critical Thinking. Paul considers critical thinking a way of life, one that is devoted to finding out the truth in a fair-minded and open way. Critical thinking is a disposition to use our critical thinking skills all the time for any subject. To Paul, critical thinking is a kind of reflective thinking that includes subjecting one's own worldview to the same kind of scrutiny and critical analysis that many of us are willing and able to do for the worldviews of those who don't think like we do. In 1981, I think it was, Paul and a few like-minded folks at Sonoma State University sponsored an international conference on critical thinking. I attended that conference and a few more at Sonoma State in the ensuing years, where I heard talks by or about several important thinkers who have come to influence my own thinking about critical thinking. (In 2012, the 32nd annual international conference on critical thinking will be held in

Berkeley, California.) One of those I heard was Robert Ennis, a philosopher of education at the University of Illinois at Urbana-Champaign, who defined critical thinking as "reasonable reflective thinking that is concerned with what to do or believe." This definition, like most definitions of critical thinking, should be seen as scaffolding (to use Paul's expression) on which to build our theories and curricula, rather than as the one and only specific goal we aim to achieve. I've moved through several definitions of critical thinking over the years, but all of them have stayed close to the core of Ennis's notion of reflective thinking that is concerned with beliefs and actions. If I had to give a definition of critical thinking, this is the one I would start with: *Critical thinking is thinking that is clear, accurate, knowledgeable, reflective, and fair in deciding what to believe or do.*

Another speaker at one of the International Conferences on Critical Thinking who influenced me profoundly was Neil Postman. Before I heard him speak, I'd read his book *Teaching as a Subversive Activity*. He later wrote a book called *Teaching as a Conserving Activity*. According to Postman, the teacher's job is not to reflect the status quo or the currently popular worldview. The teacher's job is to inspire students to think critically about that worldview. As I understood Postman, he was not advising teachers to challenge traditional algebra or geometry; nor was he advising teachers to introduce their students to crackpot scientific theories as if they constituted a serious challenge to consensus science. I don't think he would have approved of what teacher Chris Helphinstine did during his first week on the job at Sisters High School in Sisters, Oregon. The new teacher was supposed to be teaching biology, but he passed out an essay by young Earth creationist Ken Ham, who runs the website Answers in Genesis. Helphinstine also showed a PowerPoint presentation that connected evolution to eugenics experiments practiced by Nazi doctors during WWII. The new teacher said he was "hoping to encourage critical thinking in his biology class" (*The Oregonian*, 3/21/2007.) He was fired. I think Postman would have agreed that whatever else this teacher was doing, he wasn't encouraging critical thinking. He was trying to get his students to reject a consensus view in science in favor of a particular religious worldview. As I understood Postman, he was trying to get teachers in the humanities and social sciences to provide their students with alternatives to current dominating trends in those fields. He wasn't advising math and science teachers to provide junk science as an alternative to real science. If Mr. Helphinstine wanted to go outside the curriculum to teach critical thinking, he might have taught his students about consilience. Theories that have strong supportive evidence from several distinct fields are thereby strengthened. The foundation of evolutionary biology, for example, becomes stronger when facts from embryology, structural anatomy, genetics, paleontology, psychology, and other fields converge in its support.

Helphinstine is not the only one to abuse the term 'critical thinking.' In his book *The God Delusion* Richard Dawkins mentions the motto on the website of Bryan College, a Christian Bible College named after William Jennings Bryan: *Think Critically and Biblically*. I wonder what the folks at Bryan would think of a secular college that advised students: *Think Critically and Naturalistically*.

We can get a good sense of what the folks at Bryan mean by 'critical thinking' by looking at how they describe their Center for Critical Thought (now called The Bryan Institute for Critical Thought and Practice):

> Bryan College is committed to helping students develop a biblical worldview, and as part of a Christ-centered education, offers several programs toward this end. Central to the Center's work and mission is the development of exciting academic seminars in which Christian scholars who compete at the highest levels of scholarly inquiry address topics which are at the center of critical national issues. Topics include natural law, the federal judiciary system, education, taxation, science, athletics, the fine arts, and a wide range of other critical cultural concerns.
>
> Through the presentation of four seminars annually, the Center enables our academic departments on a regular multiyear basis to discuss in depth a relevant cultural issue of significance stemming from their own disciplines.

It seems that what the folks at Bryan mean by critical thinking is thinking about issues that are of critical concern to them in their mission to promote a biblical worldview and thinking about them in ways that are in accord with how they understand that worldview. They're not the only ones who understand 'critical thinking' in this way. For example, this is also what the school board in Cobb Country, Georgia, meant when it said that it was trying to encourage critical thinking by requiring a warning sticker to be placed on all biology texts:

> This textbook contains material on evolution. Evolution is a theory, not a fact, regarding the origin of living things. This material should be approached with an open mind, studied carefully and critically considered.

The school board wanted to encourage *doubting* a view they consider contrary to their understanding of the Bible. In their view, if you can encourage students to doubt a whole area of science that conflicts with a biblical worldview, you are encouraging critical thinking. This view of critical thinking is not the one that has dominated the thinking of the majority of people who teach critical thinking, who study it and write theoretical papers about it, who produce textbooks on the subject, and the like. The consensus of that group is that critical thinking requires open-mindedness. You're not encouraging an attitude of open-mindedness by telling students that what they are about to study should not be taken as fact even though the consensus of the scientific community is that it is fact. You're not encouraging open-mindedness when you advise students to think critically *and* biblically.

Furthermore, critical thinking requires a fair-minded consideration of alternative viewpoints, but the Cobb county school board was discouraging rather than encouraging fair-minded inquiry. It was mainly interested in raising doubts about evolution, which it presumably thought would enhance its own creationist beliefs. The board was not encouraging the legitimate investigation and study of various alternative evolutionary mechanisms. It was not interested in advancing inquiry but in advancing its own religious beliefs. It was specifically endorsing a false dichotomy: that any criticism of evolution implies the "only" alternative, creationism.

Using critical thinking skills to support your beliefs and to undermine opposing viewpoints is certainly legitimate, but it is a mistake to identify critical thinking with these two activities. The catalogue description for the Liberal Arts Program at Bryan College specifies that thinking critically will enable the students to "relate ideas historically and logically and compare and contrast competing views." That sounds promising, since the disposition to be open-minded enough to take seriously viewpoints that contrast with one's own is essential to being a critical thinker. But I wonder how seriously the teachers and the students at Bryan College take the viewpoints of people like Charles Darwin or Richard Dawkins.

Bryan College is located in Dayton, Tennessee, where in 1925 William Jennings Bryan successfully defended a Tennessee state law that made it illegal to teach in a state school "any theory that denies the story of the Divine Creation of man as taught in the Bible, and to teach instead that man has descended from a lower order of animals." Would the professors of biology at Bryan College encourage their students to consider that their biblical worldview might be wrong and that the theory of natural selection might be correct? This is an important question because critical thinking is much more than a set of logical skills that one uses to defend one's beliefs and refute the opposition. In fact, critical thinking is antithetical to using logical and argumentative skill to promote a particular worldview that itself is considered immune from scrutiny.

Appendix E
How to Create Your Own Pseudoscience

1. Appeal to something that most people fear or desire, things like suffering and death, or sex and longevity.

2. Make big promises about having scientific proof that you can relieve any physical illness or emotional pain, or that you can deliver "fantastic" sex or "help" people live for hundreds of years.

3. Use a lot of jargon and weasel words. Throw in words like "quantum" and "energy field" frequently. Make your product sound enormously complex, but couch all your promises with vague expressions like "may help."

4. To ward off critics who might actually know something about science, lace your promotions with references to government and business conspiracies that are keeping the truth from the general public. Make sure you remind everybody that "science doesn't know everything" and "science has been wrong before."

5. Don't be afraid to make stuff up and lie like a government leader. Even if you are prosecuted for fraud, you'll just get a lot of valuable publicity for free. The odds of you being made to suffer by a big fine or jail term are near zero. If you do have to pay a fine, change the name of your product and start over again with a few tweaks here and there in your language. You can keep doing this forever, given the kinds of things our law enforcement agencies focus on. And don't worry about the media investigating you and exposing you for a fraud. They won't bother you until you've been arrested. Even then, they'll just report that you've been charged with an "alleged" crime, which you will deny and turn in your favor by playing the persecution card.

6. Don't be cheap. Charge an exorbitant amount of money for your product. The more you charge, the more likely people, especially government procurement officers, are going to think that your product is genuine.

7. The ideal pseudoscientific product should be a hand-held device that promises eternal life, perfect health (it should detect and

cure all diseases), astounding sex (by enhancing your immune system and your personal energy flow), and can also detect bombs or golf balls with the flip of a switch.

8. Make sure you claim that you have discovered a "secret" that every other scientist in the history of the world has missed. If you're feeling especially daring, claim to have discovered a new law of nature that has scared the scientific community into trying to silence you.

9. Lace your commercials with testimonials from athletes, washed-up celebrities, and psychics. If you can get Sylvia Browne on board, do so. She has written over twenty books that have made it to *The New York Times* bestseller list. She'll be expensive, though, so if you can find someone who looks and sounds like her and will work for scale, do it.

10. Never forget that most people trust celebrities more than they trust scientists, physicians, or government agencies. Use this knowledge to your advantage.

11. Claim that the reason your work has not been published in peer-reviewed journals is because of a conspiracy to keep you silent or that the development of your product has taken all your time and money, so you haven't had the time or been able to get the funding (because of the conspiracy) to do the studies. After you make your first fortune, gather a few like-minded co-conspirators and start your own journal. You can then bypass the peer-review bias by having your peers review your many outstanding publications.

12. Don't worry about contradicting yourself. Few potential customers will notice and those who do won't care. When scientists refute you, make up ad hoc hypotheses to explain away their concerns. Don't be shy about special pleading for your product since, after all, you are a great benefactor of mankind offering them a truly unique gift. One really great device is to claim that randomized double-blind control studies don't work for your special field; that's why you don't use them.

13. Claim that some ancient civilization that can't be traced by real historians developed the technology that you re-discovered. Or claim that there are pockets of people living in

hidden villages in India or China that still use this technology and that they live for hundreds of years without the need for scientific medicine. You might even throw in the claim that skilled users of your product will be able to do some magical things like fly or at least levitate.

14. Don't be afraid to use magic tricks to deceive people, but these are rarely necessary since most people like to be deceived and won't know that you are playing on their ignorance of the placebo effect or the ideomotor effect or how to do a properly controlled test of causality.

15. Don't worry about the lack of scientific evidence for your product. People are more interested in and persuaded by testimonials than they are by scientific studies. One good story on television trumps a thousand exquisitely designed scientific studies. You might be able to get on television by connecting your product to "spirituality." You don't have to worry about defining the term, since people will make it mean whatever they want it to mean. Remember: most people love and trust stories more than they do scientific evidence. You might even enhance your own story by claiming that other scientists don't have the ability to see what you see, that the data is obscure and requires special training (that only you have) to detect what you claim to detect.

16. Most people are self-absorbed, so claim the product will reveal everything important about who they are and what's likely to happen to them in the future. Lace your testimonials with Barnum statements and claims that most people would *want* to be true for them.

17. Align yourself with the great scientists of the past who were persecuted. Galileo is a favorite here. On the other hand, you might try to distinguish yourself by taking a different route. Forget the great scientist and try to find another equally irrelevant reference, say to Andrew Wakefield, Kevin Trudeau, or Deepak Chopra.

18. If you're too lazy to create your pseudoscience from scratch, find a model to copy. For example, nobody ever went broke selling vitamin and mineral supplements. Of course you'll claim that your brand prevents cancer by stimulating the

immune system and increases IQ by increasing blood flow to the brain. Don't forget to mention that your product is organic and natural, safe for children (even if it isn't), and has been shown to improve scholastic performance by up to 80%. You might even claim that your product helps balance chi and uncoil kundalini.

19. If you're really lazy but have "charisma," you might try faith healing. You don't have to build an audience. The hopeful are waiting for you to arrive. All you have to do is show up, talk with confidence about your powers, and do a few stunts.

20. It's best to make claims that can't be disproved, but if you do make claims that can be falsified, be sure to follow the advice in point no. 12. People have many needs, but for most people the need for truth isn't one of them. Give them something to *believe* in. Promise them things like *salvation* or *eternal youth.*

21. Most people are impressed by uniforms and titles, so get a lab coat and a doctoral degree from a diploma mill. There are hundreds of diploma mills and some of them offer impressive sounding letters to place after your name for a reasonable price. You can steal a lab coat from your local hospital.

22. When proven wrong in a public demonstration of your claims, never admit defeat. Claim the study was biased, done by biased people with an agenda, or that there was one minor problem that you are going to fix as soon as more investors pour money into your pocket.

23. If you are serious about creating your own pseudoscience, you must read *How to be a Charlatan and Make Millions: A ten step programme to change your life* by Dr. Jim Williams. Here you will learn why no proposition is too stupid and that no one is immune to a good con. You'll also learn how to make your potential customers love you. You might also read Douglas Stalker and Clark Glymour's "Winning Through Pseudoscience," published in Patrick Grim's *Philosophy of Science and the Occult.*

24. On the other hand, if you are too lazy to create your own pseudoscience and selling vitamins is not to your liking, model yourself after one of the classics. Astrology and palm

reading are always good bets. Reading Tarot cards is another can't-miss favorite. The best con of all, though, is getting messages from the dead. I recommend it if you don't mind sinking to the lowest depths of depravity by desecrating sacred memories people have of loved ones. It's not as hard as you might think to master any of these arts. It's amazing how easy it is to astound people with stories about broken clocks, trips they never took, or scars they have on their knee or psyche from the trials of adolescence. Pick up a book on cold reading and you can learn to master these "arts" in a few hours. If you're going to follow my advice you're probably not the curious type, but if you ever start to wonder why people believe just about anything you tell them, study the psychology of subjective validation.

When you make your millions and become famous, don't forget those who helped you achieve your true potential and maximize your hidden talents. If we're not rewarded for our efforts, we may have to put a horrible curse on you: eternal life in heaven with people like yourself.

Sources

Chapter One

Amanzio, Martina et al. 2001. Response variability to analgesics: a role for non-specific activation of endogenous opioids. *Pain.* Feb 15;90(3):205-15.

Benedetti, Fabrizio. 2008. *Placebo Effects: Understanding the Mechanisms in Health and Disease, and The Patient's Brain: The Neuroscience behind the Doctor-Patient Relationship.* Oxford University Press.

Bausell, R. Barker. 2007. *Snake Oil Science: The Truth about Complementary and Alternative Medicine.* Oxford.

Dodes, John E. 1997. The Mysterious Placebo. *Skeptical Inquirer.* January/February.

Engel, Linda W. et al. 2002. *The Science of the Placebo - Toward an Interdisciplinary Research Agenda.* BMJ Books.

Fisher, Seymour and Roger P. Greenberg. eds. 1997. *From Placebo to Panacea: Putting Psychiatric Drugs to the Test.* John Wiley and Sons.

Hart, Carol. 1999. The Mysterious Placebo Effect. *Modern Drug Discovery.* July/August.

Hróbjartsson, Asbjørn and Peter C. Götzsche. 2001. Is the Placebo Powerless? An Analysis of Clinical Trials Comparing Placebo with No Treatment. *The New England Journal of Medicine.* Vol. 344, No. 21.

Harrington, Anne. ed. 1999. *The Placebo Effect: An Interdisciplinary Exploration.* Harvard University Press.

Kirsch, Irving , Ph.D. and Guy Sapirstein, Ph.D. 1998. Listening to Prozac but Hearing Placebo: A Meta-Analysis of Antidepressant Medication. *Prevention & Treatment*, Volume 1, June.

Price, Donald D. et al. 1999. An analysis of factors that contribute to the magnitude of placebo analgesia in an experimental paradigm. *Pain*, Volume 83, Number 2.

Price, Donald D. et al. 2005. Conditioning, expectation, and desire for relief in placebo analgesia. *Seminars in Pain Medicine.* Volume 3, Issue 1.

Shapiro, Arthur K. and Elaine Shapiro. 1997. *The Powerful Placebo: From Ancient Priest to Modern Physician.* Johns Hopkins University Press.

Smullyan, Raymond. 1986. *This Book Needs No Title: A Budget of Living Paradoxes.* Touchstone.

Chapter Two

Alcock, J. 1995. "The Belief Engine," *Skeptical Inquirer.* 19(3): 255-263.
<www.csicop.org/si/show/belief_engine/ >

Blackmore, Susan. 2003. *Consciousness: An Introduction.* Oxford University Press.

Carroll, Robert Todd. 2003. *The Skeptic's Dictionary: A Collection of Strange Beliefs, Amusing Deceptions, and Dangerous Delusions.* Wiley and Sons.

Hyman, Ray. 1999. "The Mischief-Making of Ideomotor Action," in the *Scientific Review of Alternative* Medicine 3(2):34-43.

Isaak Mark. 1998. "Problems with a Global Flood." www.talkorigins.org/faqs/faq-noahs-ark.html

Levine, Robert. 2003. *The Power of Persuasion - How We're Bought and Sold.* John Wiley & Sons.

Nyhan, Brendan and Jason Reifler. 2006. "When Corrections Fail: The persistence of political misperceptions." <www.dartmouth.edu/~nyhan/nyhan-reifler.pdf>

Zusne, Leonard and Warren H. Jones. 1989. *Anomalistic Psychology: A Study of Magical Thinking*, 2nd ed., Lawrence Erlbaum Associates.

Chapter Three

Alcock, James E. 1981. *Parapsychology: Science or Magic? – A Psychological Perspective.* Pergamon.

Gilovich, Thomas. 1993. *How We Know What Isn't So: The Fallibility of Human Reason in Everyday Life.* The Free Press.

Hirstein, William. 2004. *Brain Fiction: Self-Deception and the Riddle of Confabulation.* MIT Press.

Hyman, Ray. 1989. *The Elusive Quarry : A Scientific Appraisal of Psychical Research.* Prometheus Books.

Loftus, Elizabeth F. 1996. *Eyewitness Testimony.* Harvard University Press.

Phillips, Helen. 2006. "Mind fiction: Why your brain tells tall tales." *New Scientist.* October 7.

Schacter, Daniel L. 1996. *Searching for Memory - The Brain, the Mind, and the Past.* Basic Books.

Schacter, Daniel L. 2001. *The Seven Sins of Memory: How the Mind Forgets and Remembers.* Houghton Mifflin.

Spanos, Nicholas P. 1996. *Multiple Identities and False Memories: A Sociocognitive Perspective.* American Psychological Association.

Wiseman, Richard. 2011. *Paranormality: Why We See What Isn't There.* Macmillan.

Chapter Four

Dyson, Freeman. 2006. *The Scientist as Rebel.* The New York Review of Books.

Gingrich, Newt. 2006. "In a Free Society, Campaigns Matter: The GOP Must Give Voters a Clear Choice." <www.humanevents.com/winningthefuture.php?id=17562>

Kahane, Howard. 2001. *Logic and Contemporary Rhetoric: The Uses of Reason in Everyday Life*, 9th ed. Wadsworth Publishing Co.

Lackoff, George. 2004. *Don't Think of an Elephant! Know Your Values And Frame The Debate.* Chelsea Green Publishing.

Lutz, William. 1989. *Doublespeak.* Harper & Row.

Lutz, William. 1997. *The New Doublespeak: Why No One Knows What Anyone's Saying Anymore.* Harper Collins.

Mooney, Chris. 2004. "Beware 'Sound Science.' It's Doublespeak for Trouble." *Washington Post.* February 29.

Mooney, Chris. 2005. *The Republican War on Science.* Basic Books.

Radin, Dean. 1997. *The Conscious Universe - The Scientific Truth of Psychic Phenomena.* HarperCollins.

Rich, Frank. 2006. *The Greatest Story Ever Sold: The Decline and Fall of Truth from 9/11 to Katrina*. Penguin Press.

Rothbard, Murray N. 1987. "The Myths of Reaganomics." <www.mises.org/story/1544>

Tufte, Edward R. 1997. *Visual Explanations – Images and Quantities, Evidence and Narrative*. Graphics Press.

Chapter Five

Adams, James L. 1990. *Conceptual Blockbusting: A Guide to Better Ideas*, 3rd ed. Perseus Press.

Clarke, Richard A. 2004. *Against All Enemies*. Free Press.

Collins, Jim. 2001. *Good to Great*. HarperCollins.

Feynman, Richard P. 2001. *What Do You Care What Other People Think? Further Adventures of a Curious Character*. W. W. Norton & Co.

Griffin, Emory. 1997. *A First Look at Communication Theory*. 3rd edition. Ch. 18, "Groupthink of Irving Janis," McGraw-Hill, Inc.

Hersh, Seymour M. 2003. "Selective Intelligence," *The New Yorker*. May 12.

Janis, Irving L. 1982. *Groupthink: Psychological Studies of Policy Decisions and Fiascoes*. 2nd ed. Houghton Mifflin Company.

U.S. Secretary of State Colin Powell's Address to the U.N. Security Council <www.guardian.co.uk/world/2003/feb/05/iraq.usa >

Wikipedia. Casualties of the Iraq War. Accessed 17 October 2011. <en.wikipedia.org/wiki/Casualties_of_the_Iraq_War>

Chapter Six

Alterman, Eric. 2003. *What Liberal Media? The Truth About Bias and the News*. Basic Books.

Baldwin, Hanson W. 1963. "Managed News: Our Peacetime Censorship," *Atlantic Monthly*. April.

Barstow, David and Robin Stein. 2005. "Under Bush, a New Age of Prepackaged TV News." *New York Times*. March 13.

Carroll, Robert Todd. 2005. *Becoming a Critical Thinker - A Guide for the New Millennium*. 2nd edition. Pearson Custom Publishing.

Cockburn, Alexander and Jeffrey St. Clair. 1999. *Whiteout: The C.I.A., Drugs and the Press*. Verso.

Lakoff, George. 2004. *Don't Think of an Elephant: Know Your Values and Frame the Debate—The Essential Guide for Progressives*. Chelsea Green.

McChesney, Robert W. 1997. *Corporate Media and the Threat to Democracy*. Open Media.

McChesney, Robert W. 1999. *Rich Media, Poor Democracy: Communications Politics in Dubious Times*. University of Illinois Press.

Powell, Lewis F. 1971. "The Powell Memo." <reclaimdemocracy.org/corporate_accountability/powell_memo_lewis.html>

Radford, Benjamin. 2003. *Media Mythmakers: How Journalists, Activists, and Advertisers Mislead Us*. Prometheus Books.

Rich, Frank. 2006. *The Greatest Story Ever Sold: The Decline and Fall of Truth from 9/11 to Katrina*. Penguin Press.

Strong, Morgan. 1992. "Portions of the GULF WAR were brought to you by...the folks at Hill and Knowlton," *TV Guide*, February 22.

Chapter Seven

Angell, Marcia. M.D. 1996. *Science on Trial: The Clash of Medical Evidence and the Law in the Breast Implant Case*. W. W. Norton & Co. Inc.

Baratz, Robert. M.D., D.D.S., Ph.D. 2005. "Dubious Mercury Testing." <www.quackwatch.com/01QuackeryRelatedTopics/Tests/mercurytests.html>

Bausell, R. Barker. 2007. *op. cit.*

Blackmore, Susan J. 1987. "A report of a visit to Carl Sargent's laboratory." *Journal of the Society for Psychical Research* 54: 186P198.

Carroll, Robert Todd. 2005. *op. cit.*

Hall, Harriet M.D. 2007. "Double Blind or Double Talk? Reading Medical Research with a Skeptical Eye." *Skeptic*. Volume 13, number 3.

Hall, Harriet M.D. 2011. "Antioxidants? It's a Bit More Complicated." <www.skeptic.com/eskeptic/11-10-26/#feature>

Hyman, Ray. 1989. *op. cit.*

McCrone, John 1994. "Psychic powers: What are the odds?" *The New Scientist*. November 1994, pp. 34-38.

Satel, Sally M.D. and James Taranto. 1996. "The battle over alternative therapies," *Sacramento Bee*, January 3. First published in *The New Republic*.

Toumey, Christopher. 1996. *Conjuring Science: Scientific Symbols and Cultural Meanings in American Life*. Rutgers University Press.

Radin, Dean. 1997. *op. cit.*

Radin, Dean. 2006. *Entangled Minds: Extrasensory Experiences in a Quantum Reality*. Paraview Pocket Books.

Vedantum, Shankar. 2008. "Difficulty in Debunking Myths Rooted in the Way the Mind Works." *Skeptical Inquirer*. January/February.

Walton, Ralph G., Robert Hudak, and Ruth J. Green-Waite. 1993. "Adverse Reactions to Aspartame: Double-Blind Challenge in Patients from a Vulnerable Population." *Biological Psychiatry* v. 34, pp. 13-17.

Chapter Eight

Browne, M. Neil and Stuart M. Keeley. 2009 *Asking the Right Questions: A Guide to Critical Thinking*. Prentice Hall.

Carroll, Robert Todd. 2005. *op. cit.*

Damer, T. Edward. 2008. *Attacking Faulty Reasoning: A Practical Guide to Fallacy-Free Arguments*. 4th edition. Wadsworth Pub Co.

Moore, Brooke Noel and Richard Parker. 2008. *Critical Thinking*. McGraw Hill.

Smith, Jonathan. 2009. *Pseudoscience and Extraordinary Claims of the Paranormal: A Critical Thinker's Toolkit*. Wiley-Blackwell.

Vaughn, Lewis. 2009. *The Power of Critical Thinking: Effective Reasoning About Ordinary and Extraordinary Claims*. Oxford University Press.

Chapter Nine

Ariely, Dan. 2008. *Predictably Irrational: The Hidden Forces That Shape Our Decisions*. HarperCollins.

Bausell, R. Barker. 2007. *op. cit.*

Blackmore, Susan. 2003. *op. cit.*

Burton, Robert. 2008. *On Being Certain: Believing You Are Right Even When You're Not*. St. Martin's Press.

Carroll, Robert Todd. 2003. *op. cit.*

Chabris, Christopher and Daniel Simons. 2010. *The Invisible Gorilla: And Other Ways Our Intuitions Deceive Us*. Crown.

Dawes, Robyn M. 2003. *Everyday Irrationality: How Pseudo-Scientists, Lunatics, and the Rest of Us Systematically Fail to Think Rationally*. Westview Press.

Gilovich, Thomas. 1993. *op. cit.*

Grandin, Temple. 2010. *Thinking in Pictures: My Life with Autism*. Vintage Books.

Groopman, Jerome. M.D. 2007. *How Doctors Think*. Houghton Mifflin.

Kida, Thomas. 2006. *Don't Believe Everything You Think: The 6 Basic Mistakes We Make in Thinking*. Prometheus.

Levine, Robert. 2003. *op. cit.*

Pinker, Steven. 1999. *How the Mind Works*. W.W. Norton & Company.

Salerno, Steve. 2006. *Sham: How the Self-Help Movement Made America Helpless*. Three Rivers Press.

Shermer, Michael. 2002. *Why People Believe Weird Things: Pseudoscience, Superstition, and Other Confusions of Our Time*. Revised and enlarged edition. Holt Paperbacks.

Sutherland, Stuart. 2007. *Irrationality*. 2nd revised edition. Pinter & Martin Ltd.

Taleb, Nassim Nicholas. 2007. *The Black Swan: The Impact of the Highly Improbable*. Random House.

Van Hecke, Madeleine L. 2007. *Blind Spots: Why Smart People Do Dumb Things*. Prometheus.

Appendix A - Cell Phones, Radiation, and Cancer

Edwards, Diane D. 1988. "Cells Haywire in Electromagnetic Field?" *Science News*, v. 133, n. 14.

Lakshmikumar, S. T. 2009. "Power Line Panic and Mobile Mania." *Skeptical Inquirer.* September/October.

Livingston, James D. 1997. *Driving Force: The Natural Magic of Magnets*. Harvard University Press.

Moulder. J. E. et al. 1999. "Cell Phones and Cancer: What Is the Evidence for a Connection?" *Radiation Research,* Volume 151, Number 5, May.

Mukherjee, Siddhartha. 2011. Do Cellphones Cause Brain Cancer? *New York Times Magazine.* 12 April.

Muscat, Joshua E. et al. 2000. "Handheldellular Telephone Use and Risk of Brain Cancer," *JAMA* volume:284, December 20.

Park, Robert L. 2001. *Voodoo Science: The Road from Foolishness to Fraud.* Oxford University Press.

Pool, Robert. 1990. "Is there an EMF-cancer connection?" *Science,* v. 249, n. 4973, Sept. 7, pp. 1096-1099.

Pool, Robert. 1991. "EMF-Cancer Link Still Murky," *Nature,* v. 349, n. 6310, Feb. 14.

Richards, Bill. 1993. "Elusive Threat: Electric Utilities Brace for Cancer Lawsuits Through Risk is Unclear/ Companies Spend on Cutting Electromagnetic Fields as Lawyers Smell Blood," *The Wall Street Journal,* February 5, p. 1.

Sagan, Leonard A. 1992. "EMF Danger: Fact or Fiction?" *Safety & Health,* v. 145, n. 1 (Jan. pp. 46-49.

Trottier, Lorne. 2009. "EMF and health: A Growing Hysteria. *Skeptical Inquirer.* September/October.

Appendix C – Acupuncture, CAM, and Faith

Bausell, R. Barker. 2007. *op. cit.*

Campbell, Anthony. 2008. *Homeopathy in Perspective - Myth and Reality.* lulu.com.

Imrie, Robert. n.d. *Acupuncture: The Facts.* <drspinello.com/altmed/acuvet/acuvet.swf>

Ramey, David W. 2000. "The Scientific Evidence on Homeopathy," *Health Priorities*, Volume 12, Number 1.

Reddy, Bill. <http://www.opposingviews.com/arguments/those-who-question-scientific-basis-have-not-done-their-homework>, accessed September 2008.

About the Author

Robert Todd Carroll created the website The Skeptic's Dictionary (www.skepdic.com) in 1994, where he posts his writings about critical thinking, skepticism, and science. The dictionary part of the SD has over 700 entries from abracadabra to zombies. Other sections of the website contain essays, book reviews, a blog, and a newsletter, among other writings. *The Skeptic's Dictionary: A Collection of Strange Beliefs, Amusing Deceptions, and Dangerous Delusions* was published by John Wiley & Sons in 2003. An online Skeptic's Dictionary for Kids was published on July 22, 2011, at sd4kids.www.skepdic.com.

Carroll taught philosophy at Sacramento City College (SCC) from 1977 until his retirement in 2007. He began teaching critical thinking classes in 1975 at Lassen College where he was inspired by Howard Kahane's *Logic and Contemporary Rhetoric.* Carroll's textbook, *Becoming a Critical Thinker,* was published by Pearson in 2000. A second edition was published in 2005. While at SCC, Carroll wrote two Student Success Guides, one on Study Skills and one on Writing Skills. Both are available for free download from the SD website.

Carroll did his doctoral work in philosophy under the direction of the eminent historian of philosophical skepticism Richard H. Popkin. His dissertation, *The Common-Sense Philosophy of Religion of Bishop Edward Stillingfleet (1635-1699),* was published by Martinus Nijhoff, The Hague, in 1975.

Carroll can be contacted at sdfb@skepdic.com.

Acknowledgements

I would like to acknowledge my great debt to three men whose thinking and writing have left an indelible mark, for better or worse, on my work: Howard Kahane, Richard H. Popkin, and Martin Gardner. Kahane introduced me to what is now called "critical thinking." Popkin introduced me to skepticism. And Gardner introduced me to the world of the paranormal.

I owe much also to the many psychologists and evolutionary biologists who have made it clear that critical thinking, skepticism, and science go against the grain of many tendencies that helped our species survive and thrive.

I would like to acknowledge my debt to all the scientists who have not just *written* about science for non-scientists, but who have expressed themselves so eloquently about the glory and magic of the natural world that they have inspired millions to follow in their footsteps or, as in my case, sing their praises. These scientists are too many to name. A few whom I have found most enjoyable and inspiring are Jacob Bronowski, Carl Sagan, Stephen Jay Gould, and Richard Dawkins.

Finally, I would like to thank Dr. Harriet Hall for her constructive criticism of an earlier version of this book. I also want to thank John Renish, who has read several drafts of this book, offered many suggestions, and corrected many errors. Even though I dedicated the book to him, however, he won't take responsibility for the errors that remain. I'll have to do that. I hope there aren't too many whoppers and I hope I won't try to defend them when they're pointed out to me.

INDEX

35779379R00179

Made in the USA
Middletown, DE
15 October 2016